Intergovernmental Oceanographic Commission
technical series

10

Bruun memorial lectures

Presented at the seventh session of the IOC
Unesco, Paris, 26 October - 6 November 1971

Unesco 1972

The designations employed and the presentation of the material in this publication do not imply the expression of any opinion whatsoever on the part of the Secretariats of Unesco and IOC concerning the legal status of any country or territory, or of its authorities, or concerning the delimitations of the frontiers of any country or territory.

Published in 1972 by the United Nations Educational,
Scientific and Cultural Organization,
Place de Fontenoy, Paris 7e.
Printed in Unesco's workshops.

ISBN92-3-101014-X

Printed in France
SC. 72/XVI. 10/A

TABLE OF CONTENTS

PREFACE

Presented during the seventh session of the Intergovernmental Oceanographic Commission, this series of lectures is dedicated to the memory of the noted Danish oceanographer and first chairman of the Commission, Dr. Anton Frederick Bruun. The "Bruun Memorial Lectures" were established in accordance with Resolution 19 of the sixth session of the IOC in which the Commission proposed that important inter-session developments be summarized by speakers in the fields of solid earth studies; physical and chemical oceanography and meteorology; and marine biology. The Commission further requested Unesco to arrange for publication of the lectures and it was subsequently decided to include them in the "IOC Technical Series".

Anton Bruun was born on 14 December 1901 as the oldest son of a farmer, but a severe attack of polio in his childhood led him to follow an academic, rather than agrarian, career.

In 1926 Bruun received a Ph.D. in zoology, having several years earlier already started working for the Danish Fishery Research Institute. This association took him on cruises in the North Atlantic where he learned from such distinguished scientists as Johannes Schmidt, C.G. Johannes Petersen and Thomas Mortensen.

Of even more importance to his later activities was his participation in the Dana Expedition's circumnavigation of the world in 1928-1930, during which time he acquired further knowledge of animal life of the sea, general oceanography and techniques in oceanic research.

In the following years Bruun devoted most of his time to studies of animals from the rich Dana collections and to the publication of his treatise on the flying fishes of the Atlantic. In 1938 he was named curator at the Zoological Museum of the University of Copenhagen and later also acted as lecturer in oceanology.

From 1945-1946 he was the leader of the Atlantide Expedition to the shelf areas of West Africa. This was followed by his eminent leadership of the Galathea Expedition in 1950-1952, which concentrated on the benthic fauna below 3,000 m and undertook the first exploration of the deep-sea trenches, revealing a special fauna to which he gave the name "hadal".

The last decade of Bruun's life was devoted to international oceanography. He was actively involved in the establishment of bodies like SCOR, IACOMS, IABO, and the IOC and was elected its first president in 1961.

His untimely death a few months later, on 13 December 1961, put an end to many hopes and aspirations, but Anton Bruun will be remembered for his inspiring influence on fellow oceanographers and his scientific contribution to the knowledge of the sea which he loved so much.

THE BIOLOGY OF THE INDIAN OCEAN

G. F. Humphrey

CSIRO Marine Biochemistry Unit
The University of Sydney, Australia

The "Biology of the Indian Ocean" is a peculiarly suitable topic for a Bruun lecture to be presented before the Intergovernmental Oceanographic Commission. Anton Bruun was a biologist of a very wide and general outlook and would have known what is meant by "biology of an ocean". I do not know and I certainly would not have dared lecture on such a topic in his presence. The Bureau of our Commission, realizing how little is known of the biology of the Indian Ocean, no doubt considered that this small amount of information could easily be presented by a biochemist so ignorant of biology. In any case, our Commission is not a scientific society and therefore it would not be appropriate to give a detailed scientific account of the subject.

I think that we should start from the first main aim of the IOC, i.e. that of bringing together States willing "to participate in oceanographic programmes which require concerted action". The IOC chose the Indian Ocean as the site of its first international co-operative work and in 1961 adopted the International Indian Ocean Expedition which until then had been sponsored solely by SCOR and Unesco.

The second main aim of the IOC is "to promote scientific investigations with a view to learning more about the nature and resources of the oceans". IOC's most recent activity with regard to the IIOE was to fulfil this aim by co-sponsoring a symposium on "The Biology of the Indian Ocean" early this year at Kiel. In a volume to be published this year will be found reviews by many of the biologists who worked during the IIOE or who have subsequently studied the material and data collected by others. Further, the seven volumes to date of <u>IIOE Collected Reprints</u> contain 586 original papers, many of them on biology. There will be many other reviews and original papers, but already enough has been done to justify the wisdom of embarking on such an investigation.

The productive processes in the Indian Ocean are the biological phenomena which interest me most and I think they are also the biological aspects of most importance to man. To understand them we must make some generalizations about the distribution of the organisms therein. In this case and in all similar cases we must remember that "limits of distribution are largely a function of the sampling intensity in time and space" (Kimor, 1971). Even though I agree with such criticism of this type of work, it is still true that our main source of information on distribution of marine organisms is the comparison of such distributions with the distributions of certain other factors, whether these be simple, like temperature, or complex, like upwelling. Therefore, considerations of distribution provide not only information in their own right, but indicate special areas or phenomena for close examination. Distribution studies can never be finished; but conversely attempts at specific, separate investigations can and should commence as soon as even the most elementary distributional data are available.

PHYTOPLANKTON

We have no atlases nor even the prospect of atlases to guide us in our consideration of the distribution of phytoplankton. Most of the phytoplankton species occurring in relatively isolated areas such as the Red Sea and Persian Gulf also occur throughout the tropical Indian Ocean. In the Red Sea, <u>dinoflagellates</u> are more uniformly distributed and more abundant than diatoms; sometimes there are swarms of <u>Trichodesmium</u>. The central part of the Red Sea is relatively poor in number of phytoplankton species; diatoms and flagellates predominate. The Persian Gulf is probably poorer even than the Red Sea, both comparing unfavourably with the Arabian Gulf.

In the entire tropical zone, there is a basic complex of phytoplankton which is almost universally present. The presence of the complex in this zone and elsewhere throughout the ocean is a tribute to the different physiological properties of its members. The relative proportions of these members varies with region and is determined by metabolic responses to salinity, temperature, nutrients and vitamins. Also common in the tropical zone are several species of dinoflagellates. In general, dinoflagellates increase during weak upwelling but where there is strong upwelling there is usually also a mass development of diatoms. The subsequent preponderance of diatoms is perhaps due to their greater division rate.

The total phytoplankton biomass varies greatly with meteorological and hydrological conditions. The S.W. monsoon usually produces a several-fold increase over the N.E. monsoon season. The vertical concentration of phytoplankton is relatively independent of density structures. Only where there is a high density gradient and high water stability is the greater part of the biomass in the top 30 m (e.g. equatorial divergence region). The southern sub-tropical and western regions of the Indian Ocean have phytoplankton abundance similar to the corresponding regions of the Pacific. The eastern equatorial region is considerably poorer.

I noted above the response of dinoflagellates to even weak upwelling. It is particularly fortunate that a detailed study has been made (Taylor 1971) of the material collected on cruises of the "Anton Bruun" in the northern, central and western regions. During the S.W. monsoon distribution was rather even, except for increased concentrations in southern and western Indian waters. During the N.E. monsoon, the Andaman Sea and Bay of Bengal were particularly rich as were stations in the N. equatorial current or between it and the equatorial counter current. The band of the tropical convergence marked a diminution in abundance so that oceanic stations south of 30° S. gave very few dinoflagellates.

In the Indonesian-Australian region, Durairatnam (1964) showed that dinoflagellates predominate over diatoms. The general diminution in dinoflagellate concentration between 0 and 40° S. (Wood, 1964) produces a desert area from 10° S. to 30° S. at 98° E. to 108° E.

Very little information is available for the southern regions. South African laboratories made collections in the Agulhas area from 1961-1967 and the results of this valuable series of cruises have been summarized by de Decker (1971). Diatoms were found to be the commonest forms. The area is a very rich one and blooms were common, with several species participating. Dinoflagellates increased in number and diversity, eastward and offshore. In general, the phytoplankton composition is very similar on each side of the Cape of Good Hope. The amounts occurring on the Agulhas Bank are greatly affected by the zooplankton Thalia democratica and Doliolum denticulatum. These tunicates sometimes occupy thousands of square miles and graze heavily on the phytoplankton.

In his Antarctic studies, Sayed (1971) has confirmed that south of the sub-tropical convergence, diatoms predominate. These are usually quite different from temperate species (Wood 1964) and the dinoflagellates are represented by only a few cells of a few species.

ZOOPLANKTON

The use by many laboratories of the Indian Ocean Standard Net and the sending of large numbers of the resulting samples to the Indian Ocean Biological Centre has allowed earlier and more regular examination of material than hitherto. However, many non-standard samples were also taken. A proper summary of the distribution of zooplankton must wait until we have a synthesis of all samples. I think there are no national or international proposals for such work. Further, a summary for the whole Indian Ocean requires much more sampling. The effort needed for this might not be justified. All that we can do at the moment is to proceed as for phytoplankton and indicate some characteristics for some areas.

The northern part of the Indian Ocean is much better known than the southern and it is already clear that isolated areas such as the Red Sea and Persian Gulf are largely qualitative extensions of the Indo-Pacific region (Kimor, 1971); in quantitative terms they are impoverished both in number of species and amount of biomass.

This impoverishment is a consequence of bottom topography and hydrological conditions. The winter section by Delalo (1966) showed 11 g m^{-2} in the northern part of the Red Sea, 16 in the central part, and 23 in the southern part. The Gulf of Aden at Bab-el-Mandeb contained 78 g m^{-2}. There is recruitment from the Arabian Sea and this is of epipelagic forms brought by surface currents caused by the monsoon winds. However recruitment is insufficient to maintain a permanent high biomass. Endemic species often show morphological differences from closely related ones taken outside the Red Sea and Persian Gulf.

The central northern part of the ocean is much richer at the surface in summer (up to 200 mg m^{-3}) than in winter (up to 100 mg m^{-3}). Euphausiids predominate near the E. African coast. In general, the tropical area has only a third of its plankton biomass below 500 m. The central waters south of the equator are very poor, containing less than 50 mg m^{-3}.

This impoverishment extends south eastwards (Tranter, 1962), most regions near Australia containing less than 50 mg m^{-3}. When there is upwelling the waters between Australia and Java contain more than 100 mg m^{-3}.

In the south west of the ocean, in the Agulhas region, there is usually 100 ml m^{-3} (settled volume, de Decker, 1971) but much of this is gelatinous salps. Sometimes copepods, cladocerans and chaetognaths are in high concentration. Euphausiids were always below 0.03 ml m^{-3}.

The summary by Bogorov et al. (1969) shows that the concentration of plankton in the Indian Ocean compares more than favourably with that in the other two main oceans (Table 1). The N. tropical zone of the Indian Ocean is the richest such zone in the world. The recent death of Professor Bogorov

Table 1

MESOPLANKTON IN 0-100 METRE LAYER

	Northern			Atlantic			Pacific			Indian			Southern		
	10^6 km^2	10^6 Tons	Tons km^{-2}	10^6 km^2	10^6 Tons	Tons km^{-2}	10^6 km^2	10^6 Tons	Tons km^{-2}	10^6 km^2	10^6 Tons	Tons km^{-2}	10^6 km^2	10^6 Tons	Tons km^{-2}
Arctic	9.7	60	7	1.2	10	5									
Subarctic	2.47	70	30	1.5	20	14	2.1	70	34						
N. temperate	1.0	30	36	8.0	200	25	8.2	230	28						
N. subtropical				8.0	70	9	16.8	180	11						
N. tropical				22.8	150	7	40.3	240	6	2.5	40	18			
Equatorial				7.3	60	9	23.8	380	16	23.4	240	10			
S. tropical				17.7	70	4	46.7	310	7	20.2	80	4			
S. subtropical				7.4	30	4	13.4	50	4	7.8	40	6			
S. temperate													33	200	6
Subantarctic													24	220	9
Antarctic													12	70	6

Mesoplankton measures 0.1 to 10 mm

is a great loss to us in our consideration of marine biological resources, but I feel sure that his colleagues and students will carry on the tradition which he established.

The calculations by Cushing (1971) on the rate of plankton production showed highest values off the coasts of S. Arabia, Somalia, Malabar, and Ceylon, and in the northern part of the Bay of Bengal. The averages for all areas during the two monsoons were equal at 0.04 g C m^{-2} day^{-1}. This equality might be an artefact caused by a lack of samples from the poor central part of the ocean during the N.E. monsoon.

BENTHOS

The main factor regulating the distribution of benthos is the availability of food. The vertical food-web, from phytoplankton to benthos, reacts with special conditions created by upwelling, and produces a concentration of benthos higher in temperate than in tropical zones, and higher near the coasts than in the open ocean (Neyman et al. 1971).

The richest part of the Indian Ocean is in the N. Arabian Sea (500 g m^{-2}), the concentration falling to only a few g m^{-2} towards the south. In the Bay of Bengal, the Ganges area has up to 40 g m^{-2}, the remainder only a few g m^{-2}. The benthos of the open ocean does not reach even 1 g m^{-2}, a large central area around 90°E. being less than 0.05 g m^{-2}.

In the coastal areas of S. Africa, Australia, and Antarctica, values up to 50 g m^{-2} are common, with isolated areas up to 300 g m^{-2}. The open ocean between these areas has usually between 0.1 and 1.0 g m^{-2} (Filatova 1971).

The Indian Ocean benthos is a less self-contained fauna than those of the Atlantic and Pacific Oceans. The main part of the benthos of the N. Indian Ocean consists of Pacific species. In the S. Indian Ocean there is little Pacific influence. North of 40°S., Atlantic forms are common, while south of 40°S., Antarctic and endemic forms predominate.

FISH

There is no modern atlas of fish distribution throughout the Indian Ocean and I believe that none is contemplated. The very interesting problems associated with fish ecology in the Red Sea and Persian Gulf were reviewed by Steinitz (1971). The Persian Gulf has fewer species than the Red Sea; bathypelagic and abyssal species are almost absent. Coral fishes are the most important of the littoral fishes and there are many endemic Red Sea forms but few Persian Gulf forms.

South of 20°N., the Red Sea contains a greater amount of fish than in the north. It is probable that this is partly a consequence of a similar increase in phytoplankton, zooplankton, and other marine invertebrates and vertebrates. The marked endemic nature of the Red Sea fishes also carries through to the W. Indian Ocean.

Cohen (1971) tabulated the distribution of several genera. Most of the littoral fishes are widespread, some extending to the Pacific Ocean. The slope and abyssal fishes are less widespread. The deep-water pelagic fishes have habitats determined by temperature and salinity. The fishes of S. Africa, Australia, and the Antarctic are very different in composition and have not been sufficiently studied for generalizations to be made.

PRIMARY PRODUCTION

The rate at which carbon dioxide is assimilated by phytoplankton is a measure of the rate at which organic matter is produced. The rate is influenced by the amount of phytoplankton (usually determined as chlorophyll concentration), its physiological state, and the physical conditions (e.g. light and temperature). The IIOE Chemical-Biology Atlas being prepared by Professor Krey at Kiel will summarize and display these results.

Chlorophyll

The Red Sea shows increasing concentrations from north to south (Khemelova 1970) e.g. 0.14 mg m^{-3} at the surface in the north, 0.24 in the centre, and 0.40 in the south. The amounts in a column of water 1 m^2 in area ranged from 20 to 60 mg.

In the western Indian Ocean, chlorophyll varied greatly with season, up to 300 mg m^{-2} being found at 10°N. in the Arabian Sea (Laird et al. 1964). The maximum concentrations were usually below the surface, and their depths were usually greater in stable than in unstable waters. Water masses of high stability contained less chlorophyll.

In the eastern Indian Ocean, information is lacking from a large area of the Bay of Bengal. Concentrations are generally low or average, with no areas greater than 1 mg m^{-3}. Maxima were usually near 75 m. In central, S. African, and Antarctic areas, Ichimura and Fukushima (1963), Saijo (1963) and Saijo and Kawashima (1964) found concentrations less than 0.6 mg m^{-3}. The picture is similar for Australian waters (Humphrey 1966) except that regular sampling in one area (110°E.) revealed seasonal higher concentrations up to 1.1 mg m^{-3} (Humphrey and Kerr, 1968). In the Antarctic region, Sayed and Jitts (1971) found an average concentration of 0.28 mg m^{-3} with no values greater than 0.73. Sayed's previous extensive work allows a comparison for the Antarctic areas of three Oceans, the column figures being 20 mg m^{-2} (Indian Ocean), 13 (Pacific Ocean), and 16 (Atlantic). Corresponding figures for southern more-temperate waters are 30, 26, and 13.

Carbon assimilation

The Red Sea shows increasing activity from north to south. Khemelova (1970) obtained rates of 0.31 mg C m^{-3} hr^{-1} in the northern part, 0.71 in the central, and 3.42 in the southern. Assimilation numbers were respectively 3.91, 4.14, and 7.01 hr^{-1}, indicating a southward increase of phytoplankton efficiency. In the S. Red Sea and Gulf of Aden, radiolarians containing zooxanthellae are common; this complex often accounts for half of the assimilation (Khemelova 1967). In the western Indian Ocean there was high assimilation when and where there was high chlorophyll; rates often exceeded 1.0 mg C m^{-3} hr^{-1}. The central African coast also showed such high rates occasionally.

In the eastern Indian Ocean, there was generally less than 0.4 mg C m^{-3} hr^{-1}. The depth of maximum assimilation usually did not equal the depth of maximum chlorophyll thus suggesting that the deep phytoplankton had lost some of their biochemical capacity.

Central areas were rather low in activity but there were many patches which gave moderate values. The extensive study by Jitts (1969) along 110°E showed values up to 1.1 mg C m^{-3} hr^{-1}. The few results from southern and Antarctic regions show moderate and poor areas correlated mainly with amount of phytoplankton. The highest assimilation number found by Saijo and Kawashima (1964) was 1.6 hr^{-1}; they attribute their low values to the effects of low temperature and low light intensity on phytoplankton metabolism.

The amount of carbon assimilation can also be expressed as the amount under 1 m^2 of surface each day. The new factor introduced here is the depth to which sufficient light penetrates. Table 2 is derived from calculations made by Cushing (1971) and shows the great effect of season. For the Ocean as a whole, the mean value is 0.22 g C m^{-2} day^{-1} which is greater than for the Atlantic (0.19) or Pacific (0.13) Oceans (Moiseev, 1969).

SUMMARY OF BIOLOGY

In the Red Sea and Persian Gulf a characteristic feature is the southward increase in phytoplankton, zooplankton, benthos, fish, and primary production. The biology of the Arabian Sea and E. African Coast is greatly influenced by meteorological conditions and areas within these regions seasonally have the highest concentrations of biota and the highest activities of biological processes. We need more information on the Bay of Bengal to make useful generalizations; the influence of river discharge is probably great.

The Central Indian Ocean is somewhat impoverished, in sharp contrast to the surrounding northern, western and eastern areas. Certain parts of the western and eastern areas at certain times of the year are as biologically active as the northern area. Antarctic areas are still too little studied for generalization.

Mangrove areas are common and have high productivity, useful for artificial cultivation of molluscs, crustacea, and fish. Coral areas are also common, and being rather shallow and well-lit have a rapid turnover of productive elements. Although their rate of production is only moderate,

Table 2

ORGANIC CARBON PRODUCTION

Region	Intensity g C m^{-2} day^{-1}	Size 5° squares
Arabian upwelling	1.16 (0.23)	8 (6)
Arabian Sea	0.76 (0.12)	16 (14)
Java upwelling	0.85 (0.28)	3 (1)
E. Tropical	0.70 (0.26)	16 (12)
Equatorial	0.40 (0.15)	18 (16)
E. Africa and Mozambique	0.83 (0.42)	13 (10)
Bay of Bengal	- (0.21)	- (14)

- insufficient data

S.W. Monsoon values are followed by N.E. Monsoon in brackets it is consistent in tempo (Alverson, 1971; Tranter, 1972).

From each of the distributional studies we learn that the amounts and types of organisms vary greatly from one season to another. We learn that coastal and upwelling areas can be rich and that central areas are generally poor. We receive indications that there are exceptions - we see rich patches in desert areas.

THE FUTURE

From this knowledge and experience, I think it is clear that future work should consist of specific studies for specific purposes.

As far as purely scientific work is concerned I see little point in IOC encouraging co-ordinated expeditions to map the biological activities in the still largely-unknown areas. IOC should encourage more thoughtful and novel approaches to existing information, and specialized attacks on specific problems. Such studies have application beyond the Indian Ocean and have already been made to some extent by individuals. However, they have not yet been brought to our notice and therefore have not had an opportunity to affect our considerations at IOC.

Firstly, I refer to the work of Cushing (1971) in using IIOE data on primary production and zooplankton catches to estimate the extent of transfer from primary to secondary levels and then to estimate what amount of tertiary production might be expected. Undismayed by the weakness of the data and making imaginative allowances for that weakness, he has calculated transfer coefficients ($g\ C\ m^{-2}$. $180\ day^{-1}$ at the secondary level divided by the corresponding value at the primary level) and related them to the variation in primary production. Figure 1 shows that the coefficient decreases from low to high production areas, indicating that energy is transferred three times more efficiently in the oceanic than the coastal areas.

Although the absolute value of secondary production is still greater in coastal rather than oceanic areas, the disparity is not so great as at the primary level. If this narrowing of the gap between levels continues along the food web, and since Man harvests at the higher levels, the relative attraction of high primary production areas is weakened.

Cushing calculated tertiary production as the mean of 1% of the primary production and 10% of the secondary (Figures 2a, 2b) and it is these types of charts rather than those of simple carbon assimilation which are our best indicators of possible harvestable resources. It is by a monitoring of transfer coefficients and estimated higher-level production, that oceanographers can best aid in the continuous study of exploited resources. In analysing the species interactions during exploitation, a new powerful tool might be that of network analysis (Saila and Parrish, 1971). Such analysis has shown that whereas selective removal of species with similar physiology from eco-systems leads to lower species diversity and to instability, non-selective exploitation maintains stability. Therefore rational exploitation might be best practised as a non-monoculture. The great diversity of species in the tropical Indian Ocean might lend itself to this.

Secondly, there is the experimental work of Jeffrey (1968 and unpublished) on the interpretation of distributional studies of phytoplankton pigments. Present oceanographic survey methods produce uninterpretable information such as is shown in Figure 3 (Humphrey, 1966). The methods used in such work do not distinguish between living and dead phytoplankton and the results indicate only the maximum amount of phytoplankton present. By using chromatographic methods adapted for use at sea, it is possible not only to determine the amount of potentially photosynthetically-functional chlorophylls and carotenoids, but it is possible to obtain a semi-quantitative estimate of the amounts of each phytoplankton class and an indication as to which are being degraded (Figures 4a, 4b).

Thirdly, we have clear indications that we need to know more about the metabolic processes of the phytoplankton - the primary producers of organic matter. For example we already have general information about the distribution of dinoflagellates in many regions of the Ocean. We know of their ability to take advantage of even slight enrichment conditions. Although they are handicapped by their slow reproductive rates compared to diatoms, they have a more versatile array of metabolic characteristics on which to draw, thus allowing some species to function not only photosynthetically but also heterotrophically. In fact autotrophic and phagotrophic abilities might be of considerable importance. We need to document these functions by careful laboratory work on a range of species, and then to carry out detailed studies of their natural importance in revant ecological areas and situations. Their rôle as symbionts in relatively simple coral reef ecosystems has recently been intensively studied with novel results (Haxo, Alpha Helix, 1966, Expedition to the Great Barrier Reef).

The information on phytoplankton distribution should also be analysed mathematically. The data are often grossly incomplete but this is no excuse for not doing what can be done. The use of cluster analysis by Thorrington-Smith (1971) has allowed the delineation of phytohydrographic regions and phytoplankton associations. Her work has also shown more clearly the timing of the effects of the monsoons.

Fourthly, we should make more use of reference stations (fixed stations). These were first proposed following an informal meeting of scientists from a few of the countries participating in IIOE (Australia, France, Indonesia, Japan,

Philippines and United States of America). SCOR formally raised the question of reference stations in a circular to National Committees on 15 September 1961. It was suggested that such "stations should be decided upon in locations where ships of many laboratories could frequently and conveniently call. In addition to providing seasonal coverage at these locations, some of the observations could be used to intercalibrate the different techniques being used .... exact positions should be decided by the national committees". The stations were formally designated (Figure 5) and although data are in the world data centres, no general examination has been made and we still have no clear idea of their success. I think that these offshore points, and the laboratories at Eilat, Cochin, Djakarta, Nosy Be, Durban and Perth could form an investigational and monitoring network of considerable importance.

As far as exploitation is concerned, it is clear that the Indian Ocean is an important world resource which must be used by Man to a far greater degree. The principal use is, and will be, fisheries. This has already been realized by FAO which has established an Indian Ocean Fisheries Commission (IOFC). Already excellent progress has been made in formulating a UNDP/FAO International Indian Ocean Fishery Survey and Development Programme. The Programme Leader (Mr. John Marr) has done an outstanding job in preparing reports and programmes.

The situation needing remedial action is that around the Indian Ocean live a thousand million people with an annual protein deficit of 3 million tons. The fisheries now yield only 2.5 million tons of fish i.e. about 5% of the world total. According to Marr (1971a), within the next 10 to 20 years this could probably be increased to 20 or 25 million tons using only existing technology. To do this would require not simply the application of fisheries science or of oceanography, but also the use of economic planning and management on a large scale. Most difficult of all, it would require goodwill among nations on a global scale. It is not surprising then, that despite what was said at the beginning of the IIOE, the Expedition has not solved any fisheries problems nor has it led to increased rational exploitation of fisheries resources. What the IIOE has done is to provide an oceanographic basis for planning. Marr (1971b) has summarized this as follows.

"1. The general store of information about the Indian Ocean has been greatly increased. Some of this information (in addition to that listed in item 4 below) will eventually prove to be of relevance to fishery development, most likely in completely unexpected ways.

2. There exists a continuing interest in the Indian Ocean on the part of the world marine science community, as a result of which additions to knowledge about the Indian Ocean will continue to accrue.

3. There exists a heightened interest in marine science on the part of some, if not many, of the Indian Ocean countries, which will also result in additions to knowledge about the Indian Ocean.

4. There exists a body of information on such features, for example, as the distribution of the upwelling-high productivity areas, the depth distribution of the oxygen minimum, and the distribution and abundance of fish larvae, all of which have rather obvious relationships to fishery development."

It is quite clear that IOFC should be the forum for the management of these developments but should it be the sole forum for the science associated with them? IOC should welcome further tasks in the Indian Ocean. At least IOC could be a forum for discussion and thus promote goodwill, but there are pressing specific oceanographic problems which are peculiarly suitable for IOC attention. For example, not even 0.1% of the primary production of the area is now harvested by Man's fishing. However there is the IOFC prediction of up to a tenfold increase by present conventional means and further increase likely by new technology. From oceanographic considerations, the most likely areas for development are Somalia, S. Arabia, Malabar, Madagascar, and Java and in fact Alverson (1971) has delineated similar areas as a series of priorities for fisheries investigation (Figure 6). Again from oceanographic considerations, particularly those on the distribution of biota and on the variations in level of primary production throughout the Ocean, such large increases in exploitation would mean removal of new species at new points in the food-web. Surely IOC should assist in this development.

IOC could promote the study of the Indian Ocean as a series of ecosystems. At the moment we can see five main productive ones: enrichment areas, some oceanic areas, coral reefs, mangrove swamps, and coastal areas. The populations within these react to changes but they do not regulate to keep themselves intact. The effects of Man and Nature often determine the directions in which the systems go. This fact allows Man to hope for increasing use of the populations for his own purposes, but the increasing use brings dangers of permanent imbalance or of catastrophe. All these ecosystems need oceanographical study while our fisheries colleagues are increasing their exploitation. There will be large changes in these systems long before Man has reached the point of having to interfere with the food-web in order to bring his exploitation of the primary production up to a few per cent. It is projects in this type of fisheries oceanography which IOC should be fostering. I think there should be an FAO/IOC project of this sort. I think we should treat our advisory body ACMRR as we treat our other advisory body SCOR. We should ask ACMRR to give us specific information on the generalities I have raised and to give us proposals for joint action.

During our recent sessions we have heard much of the need for a Long-Term and Expanded

Programme of Ocean Research and for an International Decade of Ocean Exploration. Let us have these projects, but let them not be simply means of adding to our store of knowledge. Let them - make them, become projects of ocean preservation, with the continuing help of IOC for the continuing benefit of Man.

Figure 1 : Transfer cœfficients from primary to secondary production as a function of primary production

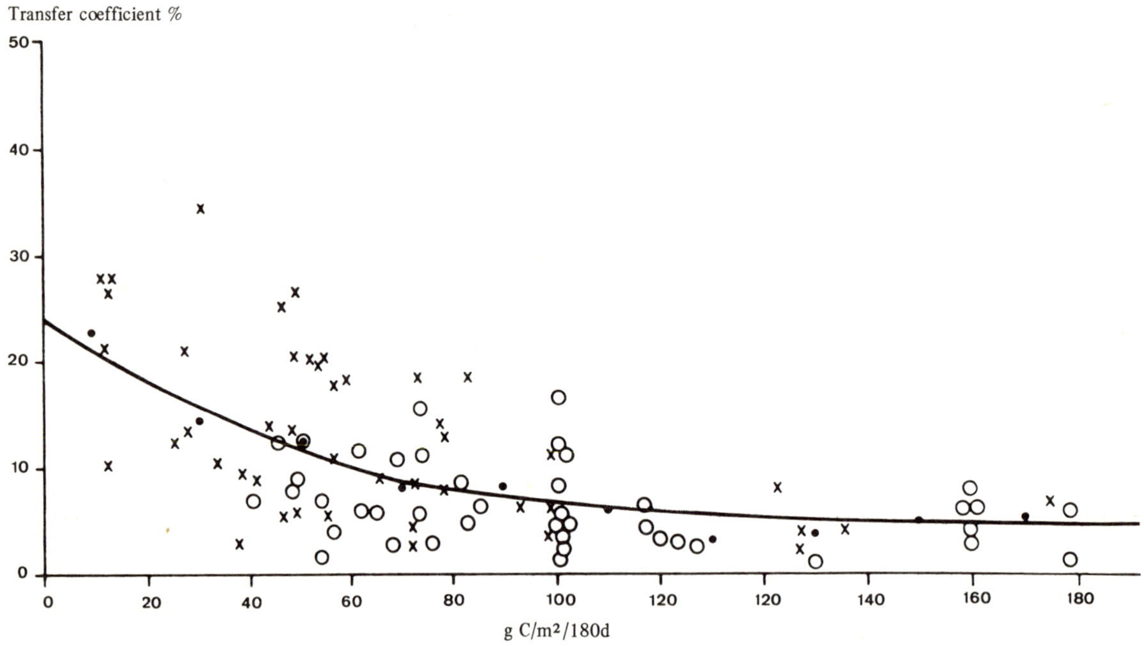

Figure 2a: The distribution of tertiary production during the northeast monsoon

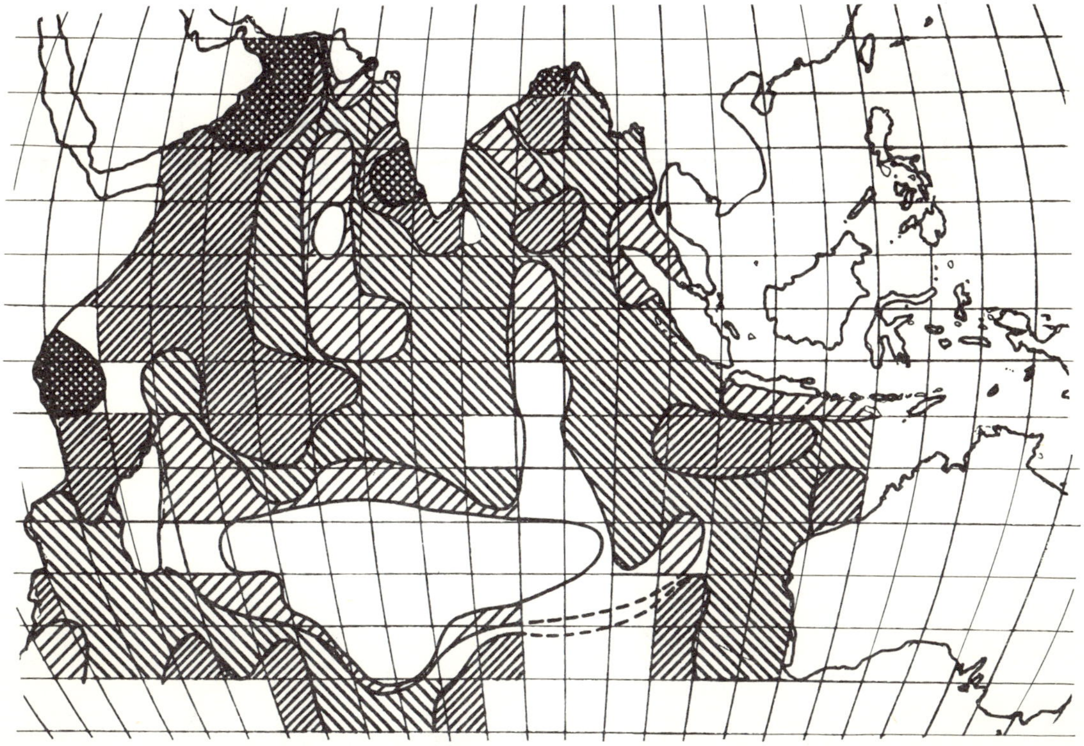

Figure 2b: The distribution of tertiary production during the south-west monsoon

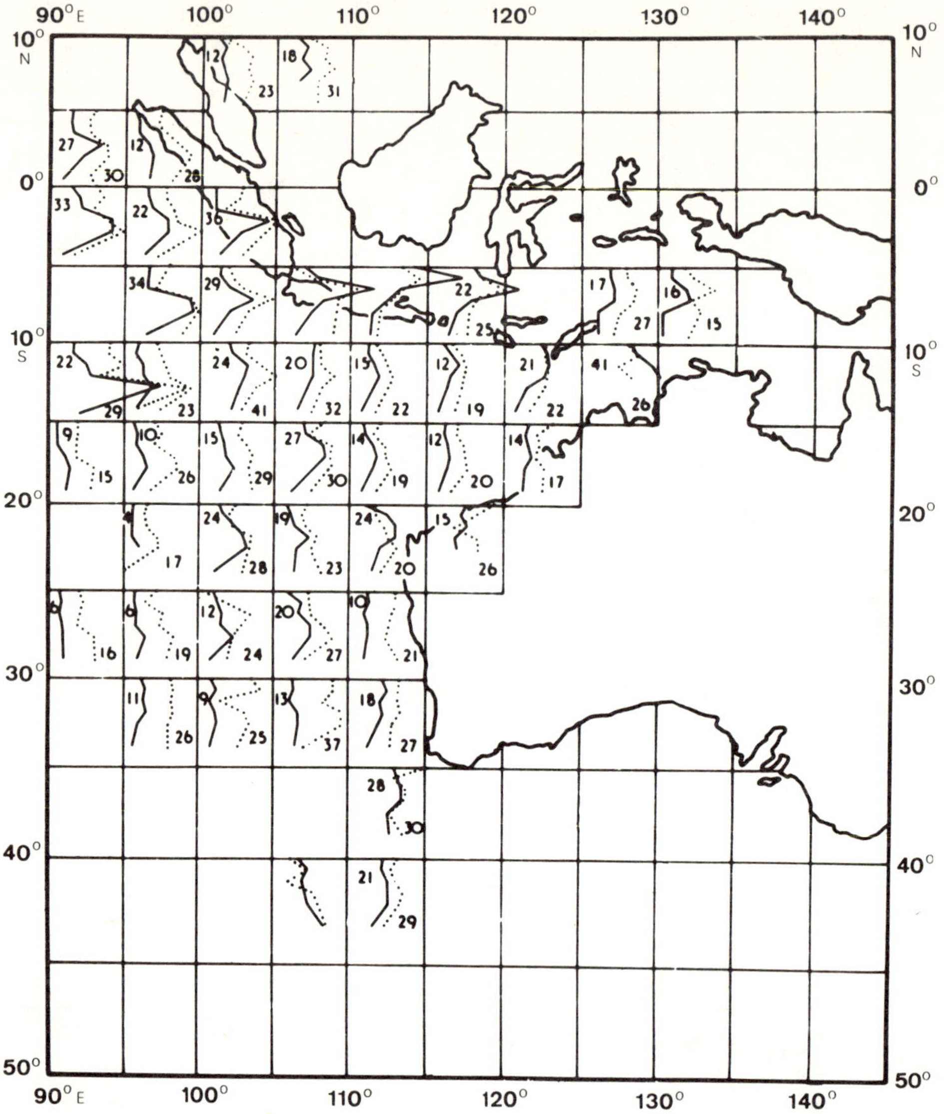

Figure 3 : Winter depth profiles of chlorophylls a (-), c (. . .) in ug/1.
The mean value at each depth for a given square is plotted

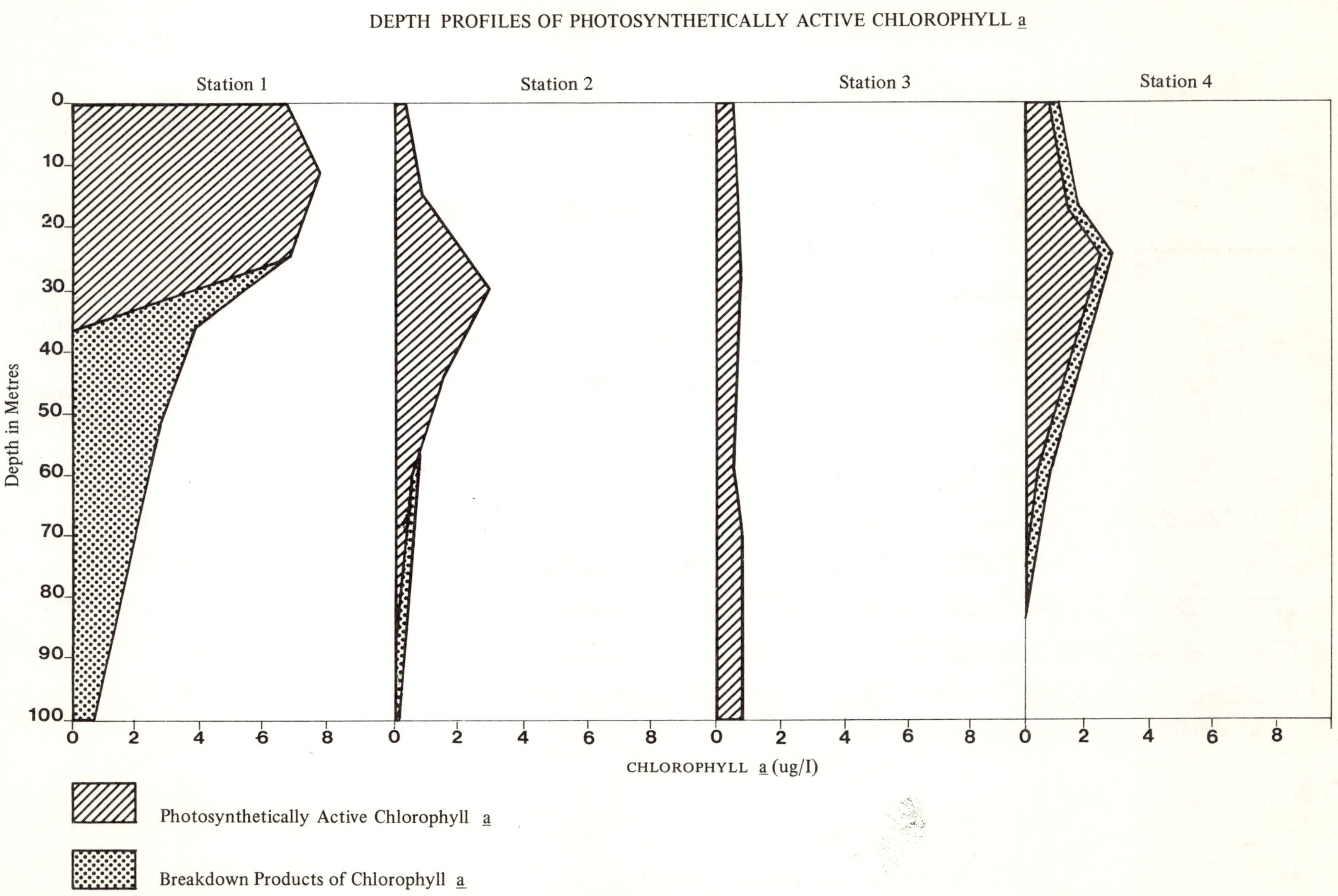

Figure 4a : Photosynthetic chlorophyll profiles

PROFILE OF PIGMENTS AT STATION 4

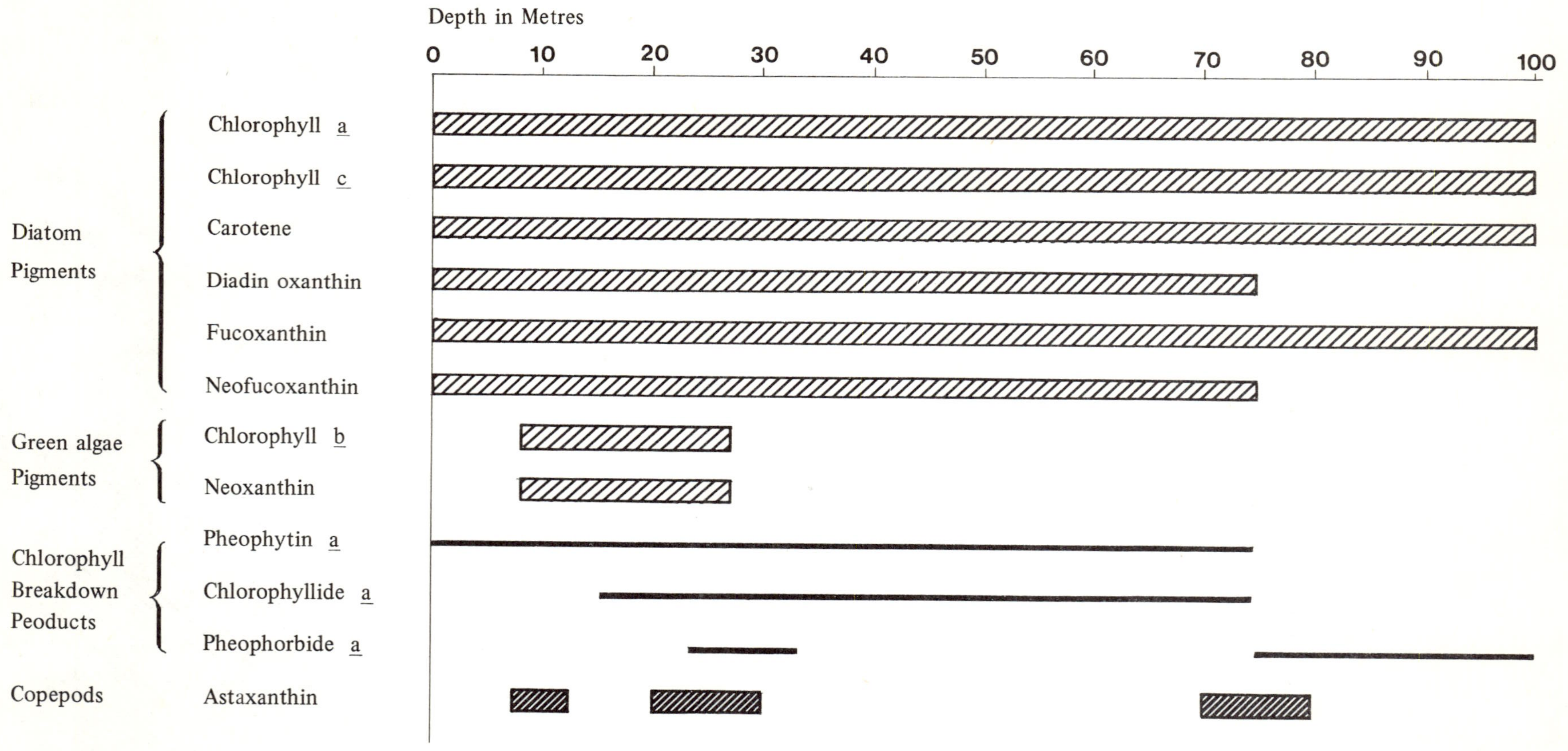

Figure 4b : Phytoplankton pigments

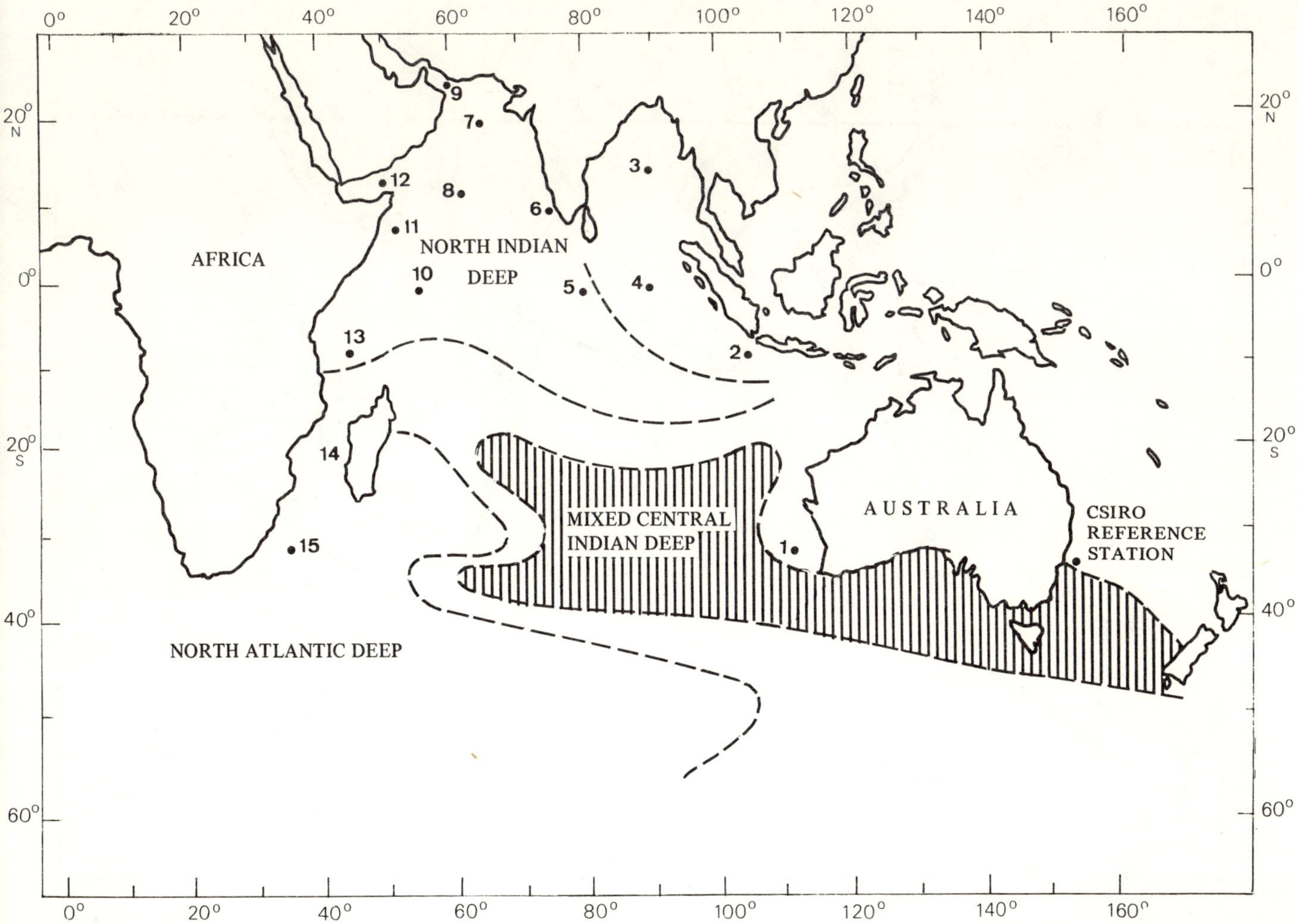

Figure 5 : Reference stations

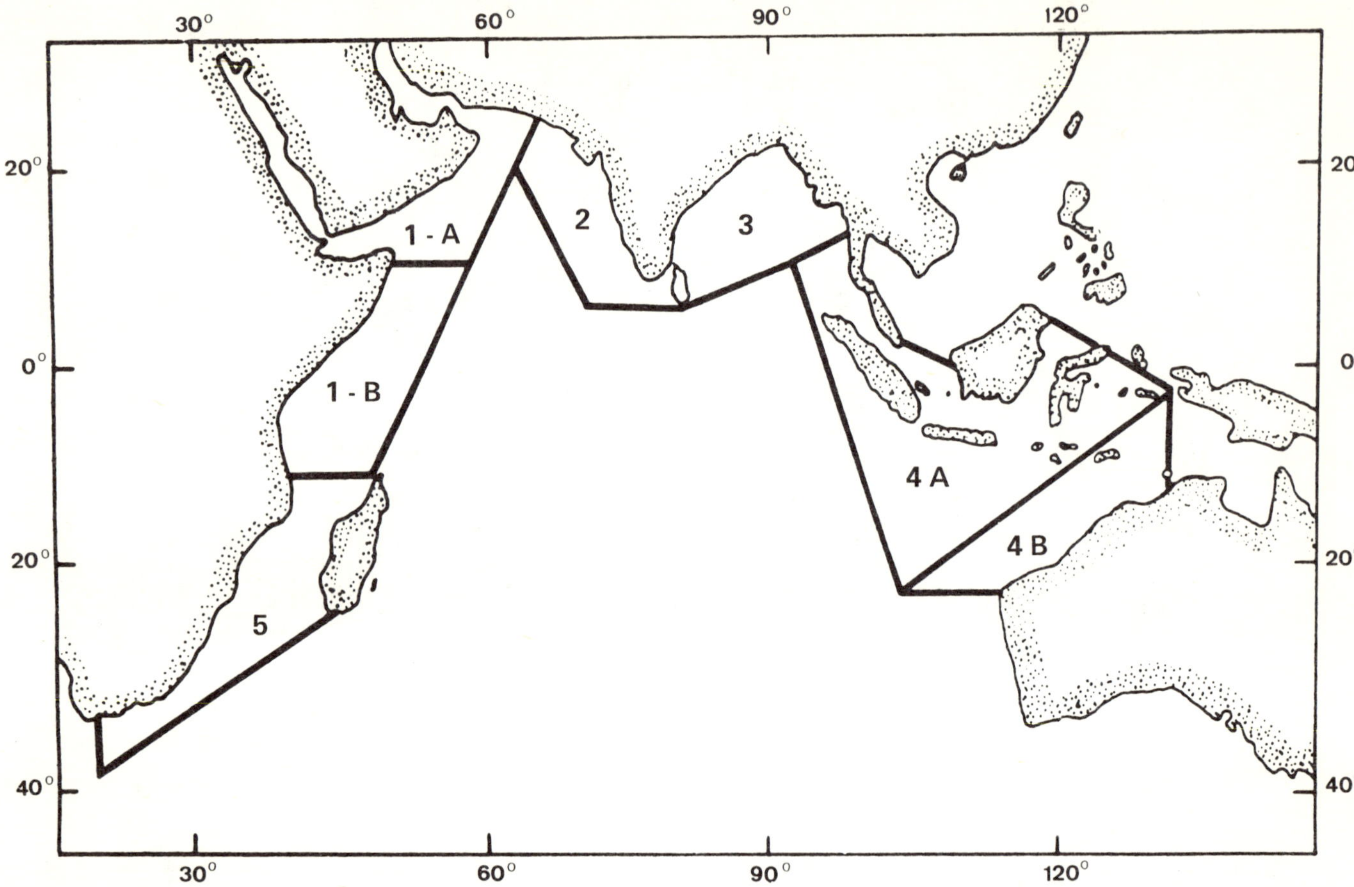

Figure 6 :   Development areas

REFERENCES

ALVERSON, D.L. (1971)
Field surveys and the survey and charting of resources. U.N. Development Programme. IOFC/DEV/71/6. ROME, FAO.

BOGOROV, V.G. et alii (1969)
The biomass of mesoplankton in the different climatic zones in the surface layer of the World Ocean. Dokl. Akad. nauk SSSR, 188(1): 209-212.

COHEN, D.M. (1971)
Zoogeography of the fishes of the Indian Ocean. To be published in: Zeitzschel, B. (ed.). The Biology of the Indian Ocean. Springer-Verlag, Berlin, (Ecological Studies).

CUSHING, D.H. (1971)
Survey of resources in the Indian Ocean and Indonesian area. U.N. Development Programme. IOFC/DEV/71/2 FAO, Rome.

de DECKER, A. (1971)
Agulhas Bank plankton. To be published in Zeitzschel, B. (ed.). The Biology of the Indian Ocean. Springer-Verlag, Berlin, (Ecological Studies).

DELALO, E.P. (1966)
Distribution of zooplankton biomass in the Red Sea and the Gulf of Aden in winter 1961-1962. Okeanol. issled. no. 15: 131-139.

DURAIRATNAM, M. (1964)
Vertical distribution of phytoplankton in an area near Cocos-Keeling Islands, Indian Ocean. Inform. Bull. Planktol. Japan, no. 11: 1-6.

FILATOVA, Z.A. (1971)
Repartition quantitative de la vie dans les oceans. Colloque International sur l'Exploration des Oceans. Bordeaux, France, March 1971. Theme II, Marketing of marine products and development of marine cultivation, 2: G 1-00.

HUMPHREY, G.F. (1966)
The concentration of chlorophylls a and c in the south east Indian Ocean. Aust. J. mar. Freshwat. Res., 17: 135-145.

HUMPHREY, G.F. and J.D. KERR (1969)
Seasonal variations in the Indian Ocean along 110°E. III. Chlorophylls a and c. Aust. J. mar. Freshwat. Res. 20: 55-64.

ICHIMURA, S. and H. FUKUSHIMA (1963)
On the chlorophyll content in the surface water of the Indian and Antarctic Oceans. Bot. Mag., Tokyo, 76: 395-399.

JEFFREY, S.W. (1968)
Quantitative thin layer chromatography of chlorophylls and carotenoids from marine algae. Biochim. biophys. Acta, 162(2): 271-285.

JITTS, H.R. (1969)
Seasonal variations in the Indian Ocean along 110°E. IV. Primary production. Aust. J. Mar. Freshwat. Res. 20: 65-75.

KHMELEVA, N.N. (1967)
Rôle of radiolarians in the estimation of the primary production in the Red Sea and Gulf of Aden. Dokl. Akad. nauk SSSR, 172(6): 1430-1433.

KHMELEVA, N.N. (1970)
On the primary production of the Red Sea and the Gulf of Aden. Biology of the Sea, 21: 107-133.

KIMOR, B. (1971)
Plankton relations of the Red Sea, Persian Gulf and Arabian Sea. To be published in Zeitzschel, B. (ed.). The Biology of the Indian Ocean. Springer-Verlag, Berlin, (Ecological Studies).

LAIRD, J., B.B. BREIVOGEL and C.S. YENTSCH (1964)
The distribution of chlorophyll in the western Indian Ocean during the southwest monsoon period.

30 July - 12 November 1963. Woods Hole Oceanogr. Inst. Tech. Report No. 64-33.

MARR, J.C. (1971a)
Indian Ocean fishery development. To be published in Zeitzschel, B. (ed.), The biology of the Indian Ocean. Springer-Verlag, Berlin, (Ecological Studies)

MARR, J.C., D.K. GHOSH, G. PONTECORVO, B.J. ROTHSCHILD and A.R. TUSSING (1971b).
A plan for fishery development in the Indian Ocean. U.N. Development Programme. IOFC/DEV/71/1 FAO, Rome.

MOISEEV, P.A. (1969)
Biologicheskie resursy mirovogo okeana. Izdatelstvo Pishchevaya Promyschchlennost, Moscow. 339 p.

NEYMAN, A.A., M.N. SOKOLOVA, N.G. VINOGRADOVA, Th. A. PASTERNAK (1971)
Some patterns of the distribution of the bottom fauna in the Indian Ocean. To be published in Zeitzschel, B. (ed.). The biology of the Indian Ocean. Springer-Verlag, Berlin, (Ecological Studies).

SAIJO, Y. (1963)
Technique of primary production measurement used on the Indian Ocean cruise of the Umitaka Maru, December 1962 - January 1963, and suggestions on their improvement on the next cruise. Inf. Bull. Planktol. Japan, 10: 17-22.

SAIJO, Y. and KAWASHIMA, T. (1964)
Primary production in the Antarctic Ocean. J. oceanogr. Soc. Japan, 19: 190-196.

SAILA, S.B. and J.D. PARRISH (1971)
Exploitation effects upon interspecific relationships in marine ecosystems. Colloque international sur l'Exploitation des Oceans. Theme II. Marketing of marine products and development of marine cultivation. 1: G1-09.

SAYED, Z. El-SAYED and H.R. JITTS. (1971)
Phytoplankton production in the south eastern Indian Ocean. To be published in Zeitzschel, B. (ed.). The biology of the Indian Ocean. Springer-Verlag, Berlin, (Ecological Studies).

STEINITZ, H. (1971).
Fish ecology of the Red Sea and Persian Gulf. To be published in Zeitzschel, B. (ed.). The biology of the Indian Ocean. Springer-Verlag, Berlin, (Ecological Studies).

TAYLOR, F.J.R. (1971)
General features of dinoflagellate material collected by the "Anton Bruun" during the International Indian Ocean Expedition. To be published in Zeitzschel, B. (ed.). The biology of the Indian Ocean. Springer-Verlag, Berlin, (Ecological Studies).

THORRINGTON-SMITH, M. (1971)
West Indian Ocean phytoplankton: a numerical investigation of phytohydrographic regions and their characteristic phytoplankton associations. Marine Biology, 9(2): 115-137.

TRANTER, D.J. (1962)
Zooplankton abundance in Australian waters. Aust. J. mar. Freshwat. Res., 13: 106-142

TRANTER, D.J. (in press)
Marine biology. In: A review of the natural resources of humid tropical Asia. Paris, Unesco. (Natural Resources Research series).

WOOD, E.J.F. (1964)
Studies in microbial ecology of the Australasian region, Nova Hedwigia, 8 (1 & 2): 5-54. 8 (3 & 4): 453-568.

RECENT METEOROLOGICAL RESULTS FROM THE INTERNATIONAL INDIAN OCEAN EXPEDITION*

C. S. Ramage

University of Hawaii, Honolulu, Hawaii

INTRODUCTION

The impetus given Indian Ocean meteorology by the International Indian Ocean Expedition continues to be felt. Published data compilations and analyses are encouraging individuals not originally connected with the IIOE to study Indian Ocean phenomena.

During the IIOE meteorologists established a network of radiation stations (Portman and Ryznar, 1971) and measured ocean-atmosphere exchanges from a floating, instrumented mast (Badgley et al., 1971). Apart from these activities and reconnaissances by meteorological research aircraft, the meteorologists "went along for the ride". We compiled observations from existing land stations, merchant ships and commercial aircraft and so increased our knowledge of Indian Ocean climate. We made as many measurements as possible from oceanographic research vessels sailing on cruises designed for non-meteorological purposes. We were philosophical about this, for our ignorance was often too great to allow us to specify ways of reducing it. As it turned out, the oceanographers' emphasis on the Arabian Sea during summer suited us fine and, as I shall point out later, oceanographic and meteorological observations have been combined to develop a physically consistent picture of the summer atmospheric circulation over that sea. Regrettably, the Bay of Bengal was not so well surveyed; had it been, the similarities and differences between circulations over the Bay and the Arabian Sea would surely have illuminated our understanding of both.

Although the pattern of cruises failed to approach the uniformity specified by Wüst (1960), weather satellite pictures have very usefully supplemented other observations. Since the IIOE, a detailed cloud climatology (Sadler and Harris, 1970) based on five years of continuous surveillance has helped fill in many of the gaps.

With the explosive expansion of computer applications to meteorology, the Indian Ocean has not been neglected; in fact, the excellent data compilations of the IIOE have formed the basis for evaluating various diagnostic models of the monsoon circulations and some of their components. As we shall see, our insight into complex cause and effect interactions has been thereby deepened.

In the remaining sections I discuss climatological atlases and the derived fields of surface wind divergence and surface wind stress curl. Then follows a section dealing with features of the large-scale summer circulation over the Arabian Sea and the rôle of the Himalayas and Tibet as revealed by numerical modelling. Evidence for the effect of valley winds on the sea is presented next. Finally, using radiation measurements, cloudiness as observed by weather satellites, and measurements of evaporation, I have attempted to evaluate some of the empirical formulations which relate standard meteorological observations to the ocean-surface heat budget.

CLIMATOLOGY

IIOE Meteorological Atlas. One of the prime purposes of the IIOE was to collect data for a series of atlases (Intergovernmental Oceanographic Commission, 1965). A two-volume Meteorological Atlas is being published by the National Science Foundation, Washington, D.C.

Projection (Mercator's) and scale (1: 40 million) conform to World Meteorological Organization specifications for meteorological atlases.

Volume I (Ramage, Miller and Jefferies, 1971) describes the surface climate of the Indian Ocean as determined from 194,000 ship observations made during 1963 and 1964 when expedition activity was at its height. From the observations, averages - by individual month and by 5-degree latitude-longitude square - of wind, pressure, air- and sea-surface temperatures, vapour pressure, clouds, precipitation, and heat exchange, together with analyses are depicted on 144 charts.

Meteorologists and oceanographers will be

* Contribution No. 71-4 of the Department of Meteorology, University of Hawaii.

able to use the volume as a background to detailed studies, enabling them to relate physical and biological distributions to the monthly climate. Also, by comparing the charts with long-term meteorological atlases, they will be able to determine significant departures from the norm.

Volume II (Ramage and Raman, 1971), which describes the upper-air climate of the Indian Ocean and its neighbouring continents, incorporates information from before, during and after the expedition period. Monthly long-term averages of resultant wind, wind steadiness, pressure-height, temperature, and dew-point are presented on two sets of charts - i.e., constant pressure level charts for 850, 700, 500, 300, 200 and 100 mb (approximately 1.5, 3.1, 5.8, 9.6, 12.4 and 16.5 km above m.s.l., respectively) covering the area between 45°N and 50°S and 20°E and 155°E, and meridional cross-section charts for the longitudes 30°E, 73°E, 110°E and 140°E. On the constant pressure-level charts, analysis consists of streamlines and isotachs; on the meridional charts isotachs of the zonal wind component and isentropes are shown.

In addition to averages computed from sounding-baloon observations, averages of flying-level winds measured from aircraft provide significant detail over the open ocean.

Meteorologists, climatologists and airline operators should be among those finding uses for this volume. From the plotted data they will be able to construct different sets of isopleths.

Fields derived from mean resultant surface winds. What is still the finest Indian Ocean surface climatological atlas was prepared in the Royal Netherlands Meteorological Institute (van Duijnen Montijn, 1952). It depicts, for each 2-degree square, mean resultant surface winds computed from merchant ship observations. Unfortunately, because of great areal variability in numbers of observations, "noise" in derived fields, such as divergence, almost conceals large-scale patterns (van Dijk, 1956). Hantel partly overcame this handicap. Data gaps were filled by linear interpolation from neighbouring squares with many observations, then the zonal and meridional components of the wind were filtered to remove wavelengths of less than 900 km. The amended data were then processed to yield realistic fields of surface wind divergence (Hantel, 1971) (see Figure 7) and surface wind stress curl (Hantel, 1970). Of course, important small-scale features may have been concealed in the process.

ATMOSPHERIC CIRCULATION OVER THE ARABIAN SEA IN SUMMER

To the delight of meteorologists oceanographers concentrated on this region and on this season during the IIOE. Here the greatest gradients of temperature and weather are observed providing the best chances of scattered measurements leading to useful physical insights.

The atmospheric circulation is strong and persistent (Figure 8). Below 1 - 1.5 km. winds sweep across the equator in the west and veer from south westerly to west-south westerly toward western India. Strong upwelling is induced off Somalia and Arabia and the consequent increase in temperature gradient there induces a low-level jet stream in which speeds reach 25 m sec^{-1} (Bunker, 1967a). Above 7 km., the circulation is almost reversed; strong east-north easterly winds often exceed 30 m sec^{-1}. Between 2 and 5 km., winds from a westerly quarter predominate.

West of a line extending from Karachi to the equator at 50°E, very little rain falls; east of this line, and particularly along and off the coast of western India, rain is common and frequently heavy. Distribution of the divergence of the mean winds fits this pattern. Near the surface (Figure 7) flow is generally divergent (implying sinking and stabilization) in the dry area and convergent (implying rising and destabilization) in the wet area (Flohn et al., 1968). But there is a paradox. Surface air reaching the coast of Baluchistan is moisture-laden after a long sea-passage and in fact possesses a slightly higher vapour pressure than surface air at Bombay. Also, surface convergence is at least as great and the near-coastal ranges are at least as high. Why then does Bombay receive an average of 1,700 mm of rain from June through September while the Baluchistan coast receives less than 25mm?

Effect of Himalayas and Tibet. The answer lies in the rôle of the Himalayan-Tibetan massif (Ramage, 1969). During summer, central and south eastern Tibet acting as a high-level radiational heat source, and the Himalayas, by enhancing precipitation and the consequent release of latent heat, create a temperature maximum in the high troposphere. Temperature gradients are consequently larger toward the south and south west than in other directions and so a speed maximum develops in the upper-level easterlies, with downwind convergence and sinking motion over the western and north western Arabian Sea and the bordering deserts and divergence and rising motion over the south eastern Arabian Sea and most of India. Surface air converges into a heat low over west Pakistan but subsidence induced by Himalaya-Tibet so restricts its vertical motion that precipitating clouds very rarely form.

Murakami et al. (1969) applied the equations of motion to an eight-layered atmosphere to develop a two-dimensional numerical model of the summer monsoon extending along 80°E from the equator to the North Pole. Neither disturbances nor trans-equatorial flow were permitted. Effects of radiational and condensation heating were allowed for; within the surface turbulence layer, vertical fluxes of momentum, heat and water vapour were assumed to be constant and were computed using empirical formulae.

The investigators started their computations with a completely calm and dry atmosphere possessing the Standard Atmosphere vertical temperature distribution. In their latest experiment they included evaporation at the earth's surface, condensation in the atmosphere, and mountain effect. Sea-surface temperature was kept constant at 27°C and the following assumptions were made: land albedo - 15 per cent; surface relative humidity over the continent - 60 per cent; surface relative humidity over the ocean - 100 per cent; and condensation relative humidity - 100 per cent. Integration carried out over a period of 80 days using 10-minute time-steps, determined a zonal wind profile. Comparison with the long-term mean for 80°E (Figure 9) reveals only slight discrepancies, largely resulting from the assumption of no trans-equatorial flow. Values of speed maxima agree.

This is particularly significant since in earlier experiments which did not include the mountain effect, the westerly maximum was computed to be less than 5 m sec^{-1} and the easterly maximum less than 10 m sec^{-1}, comparable to averages along across sections east of the Himalayas. Thus, the model strikingly confirms the potent effect of Himalaya-Tibet on the circulation.

Subsidence. Research vessels and aircraft made 172 summertime radiosoundings of temperature and moisture over the Arabian Sea during the IIOE. These reveal further details of the vertical circulation, since the lower limit of subsiding air is marked by a temperature inversion (temperature increasing with height) and a sharp decrease in moisture through the inversion. Figure 10 shows that the inversion is lowest and most intense where fine weather prevails and absent where the weather is predominantly bad.

Frequently, over the central Arabian Sea, middle and high clouds are observed with their bases 1 - 10 km above the inversion-limited tops of low clouds (Bunker and Chaffee, 1970). The former obviously cannot originate from the latter but must be spread westward from the tops of deep cloud masses in the rising part of the circulation over India. Soundings by *Discovery*, *Argo* and *Atlantis II* in the region of strong upwelling showed that temperatures increased continuously from the cold sea surface up through the subsidence inversion. Thus, although surface air moving from the south was rapidly cooled to its dew point by the cold water, fog never formed because the heat abstracted by the sea was continuously replaced from the reservoir of warm subsiding air above (Ramage, 1968).

Some meteorologists, members of the "air mass school", account for the high temperatures above the inversion not in terms of compressional heating due to subsidence but as reflecting the presence of tropical continental air which moves eastward after being heated over the deserts of Africa and Arabia (Desai, 1967). Were this so, one would expect the air to cool as it moved farther from its heat source. However, the middle tropospheric westerlies are, level for level, warmer over the Arabian Sea than over Africa, a difference that can be explained only by subsidence.

Radon as a tracer. Rama (1968) measured concentrations of radon (disintegrations per minute per cubic metre) at deck level during four summertime cruises in the Arabian Sea (Figure 11). Since radon is almost entirely continental in origin, concentrations > 50 dpm m^{-3} are found in air of continental origin; concentrations of < 5 dpm m^{-3} are found in maritime air; and intermediate concentrations probably result from the mixing of maritime and continental air. The three régimes are readily recognizable in Figure 11. Near the African coast at 5°S and in the Gulf of Aden and the Red Sea vigorous, diurnally-varying local wind systems (Flohn, 1970) continuously cycle radon-enriched air. Farther east between 5°S and 15°N the subsidence inversion seals off the continental westerlies (Figures 8 and 10) and the low concentrations are typically maritime. East and north east of this zone the continental westerlies are usually found below the subsidence inversion, vertical mixing is not inhibited and surface concentrations reflect mixing of continental and maritime air.

ATMOSPHERIC AND OCEANIC REACTIONS TO VALLEY WINDS

The Western Ghats of India locally distort the north east monsoon in at least two ways. The winds up to 100 km offshore are constrained to flow parallel to this nearly continuous 1, 300 km. long mountain chain. However, just below 11°N the range is broken by a major valley, the Palghat Gap, through which winds are channelled from the east. In studying weather satellite photographs made during February 1964, Bunker (1967b) identified meso-scale cloud systems off the coast. These were often found between 11°N and 12°N and somewhat less frequently between 8°N and 9°N. With the aid of detailed wind, temperature and moisture measurements made during a Woods Hole Oceanographic Research Institution meteorological flight on 19 February 1964, Bunker hypothesized that the northern cloud lines probably developed as a result of convergence (and consequent rising motion, cooling and condensation) between the valley wind and the wind blowing parallel to the Western Ghats.

Such a topographically anchored feature of the winter monsoon might be expected to persist long enough to affect the near-shore surface layers of the ocean. With this possibility in mind I inspected the records of surface temperature and surface salinity made on board *Meteor* during a coastal cruise from Cochin to 15°N during February 1965 (Figure 12). Although no obviously orographic clouds appeared on weather satellite photographs for this period, the air might well have been too dry for the lifting condensation level to have been reached. Figure 12 reveals three rather distinct zones: in Zones 1 and 3 temperature is relatively low and salinity relatively high, whereas in Zone 2 temperature is relatively high and salinity relatively low (Table 3).

The boundary between Zone 1 and Zone 2 probably coincides with the northern limit of easterly winds blowing out of the Palghat Gap. Thus, if the winds did affect the sea, Zone 1 should experience upwelling resulting from a combination of winds blowing directly offshore and positive curl of the wind stress. In Zone 2, upwelling would be inhibited by negative curl of the wind stress (valley easterlies and coastal northerlies farther north). In Zone 3, north-northwest winds blowing parallel to the coast would favour upwelling. These small-scale details are not resolved in Hantel's (1970) charts of mean monthly surface wind stress curl.

Over the eastern Arabian Sea during the north east monsoon, salinity increases with depth. In regions of upwelling relatively higher surface salinities would accompany relatively lower surface temperatures, as in Zones 1 and 3.

The Palghat Gap winds apparently produce measurable meteorological and oceanographic effects. They are similar to but much weaker than those produced by strong winter northerlies on the Gulf of Tehuantepec (Roden, 1961).

COMPONENTS OF OCEAN-SURFACE HEAT EXCHANGE

Solar radiation. Solar and atmospheric radiation measurements were made by several groups. Two have published their results.

A team led by Donald J. Portman (Portman and Ryznar, 1971) of the University of Michigan established stations on seven Indian Ocean islands and at six coastal locations in India and Africa. The data obtained during 1963 and 1964 were analysed. Information on cloudiness and water-surface temperature was used in computing the remaining components of the net exchange of thermal radiation to arrive at monthly averages of net radiative exchange for the two years.

Net radiative exchange for each month of 1963 and 1964 was also computed for the meteorological atlas, using Budyko's (1956) tables for total radiation from a clear sky. Figure 13 compares Budyko's values with those determined by Portman and Ryznar from averages of daily integrated solar radiation measured by Eppley pyrheliometers. Except between the equator and 5° - 10°S, Portman and Ryznar's values are significantly lower. They explain this as follows: first, Budyko's results were based largely on measurements made where average values of precipitable water could have been less than for the Indian Ocean region. This appears plausible, for the difference between the two sets of values are least near the equator, where longitudinal variations in precipitable water are also least, and greatest near 25° latitude where longitudinal differences (between deserts and oceans) are greatest. Second, just before the Indian Ocean network was established, Mt. Agung in Bali erupted. The large amount of volcanic material it discharged high into the atmosphere almost certainly attenuated the solar radiation.

Total solar radiation was measured on two Australian cruises to the eastern Indian Ocean during 1962 (Deacon and Stevenson, 1968), before Mt. Agung erupted. Characteristics of the instrument, a Moll-Gorczynski solarimeter, resemble those of the Eppley pyrheliometer (Fuquay and Buettner, 1957). From reported spot readings on clear days, I calculated average daily values for four locations (Table 4 and Figure 13). They do not reveal any eruption effect, differing only slightly from Portman and Ryznar's values, which might well replace Budyko's data in any future computations of Indian Ocean surface heat exchange.

Cloudiness figures importantly in all computations of net ocean-surface heat exchange. However, cloudiness is seldom accurately estimated by an observer on shipboard and is often unrepresentative of conditions just below the horizon. These shortcomings might cancel out in long-term averaging but could seriously reduce the value of inter-annual comparisons.

Weather satellites depict the whole field of cloudiness. Sadler (1969) and Sadler and Harris (1970) have computed monthly means for the tropics derived from five years of once-daily (about local noon) weather satellite passes. Cloudiness was averaged for each $2^1/2$ degree latitude-longitude square between 30°N and 30°S. Figure 14 compares these averages with long-term means based on ship reports (U.S. Weather Bureau, 1957). The curves are encouragingly alike. The maximum difference shown would lead to a difference in computed net heat exchange of less than 10 per cent of the incoming clear-sky radiation. Now is perhaps the time to recompute the effect of cloudiness on the radiation components using satellite-determined values of cloudiness. Inter-annual or shorter period variations would be readily calculable since weather satellites will continue to measure cloudiness with unequalled accuracy.

Evaporational cooling. Evaporational cooling (Q_e) is of the same magnitude as net heating due to radiation. In the IIOE Meteorological Atlas it was computed from meteorological data by the empirical formula $Q_e = 5.17 (e_s - e_a) V$ based on Sverdrup (1951). Q_e is expressed in ly day^{-1}, e_s is the saturation vapour pressure in mb of air at the sea surface temperature, e_a is the vapour pressure in mb of the air at 8 m above the surface and V is the wind speed in m sec^{-1} also at 8 m above the surface.

During February and March 1964, University of Washington scientists measured sea surface temperature, and wind, temperature and moisture content at six levels up to 8.15m on an instrumented floating mast at four locations in the north east Arabian Sea (Badgley et al., 1971). Table 5 compares turbulent fluxes of latent heat (evaporational cooling) determined from analysing the vertical profiles of moisture content measured on the mast with evaporational cooling calculated using Sverdrup's formula.

During these same periods, the Woods Hole Oceanographic Institution's research aircraft flew nine rectangular sorties around the instrumented mast. From measurements of humidity and wind velocity made on the aircraft and by dropsondes released from it, evaporation from the sea surface was calculated, being the difference between the horizontal fluxes of water vapour into and out of the rectangles (Fleagle *et al.*, 1967). Over the whole period of the experiment, evaporational cooling calculated using Sverdrup's formula averaged 114 per cent of the value determined from the mast profiles and 91 per cent of the value determined from the horizontal fluxes. Since the formula gives results close to the average of the two measurement methods, there is little reason to modify it, at least for wind speeds of less than 6 m sec^{-1}, when no more than 1/200 of the sea surface is covered by white caps (Monahan, 1971). However, it would be unwise to extrapolate this conclusion to the turbulent conditions associated with stronger winds.

CONCLUSION

Despite the patchy nature of Indian Ocean meteorological observations (except for those from weather satellites), information garnered during and after the IIOE has improved our knowledge of climatology, the summer atmospheric circulation over the Arabian Sea and the components of ocean-surface heat exchange.

Already, qualitative physical insights have been powerfully enhanced by limited numerical diagnostic studies; future understanding of complex ocean-atmosphere and trans-equatorial interactions will depend even more on successful application of computer techniques.

Acknowledgement. I thank Dr. Peter H. Koske of the Institute of Physical Chemistry of the University of Kiel for providing me with surface temperature and salinity data from the *Meteor* cruise.

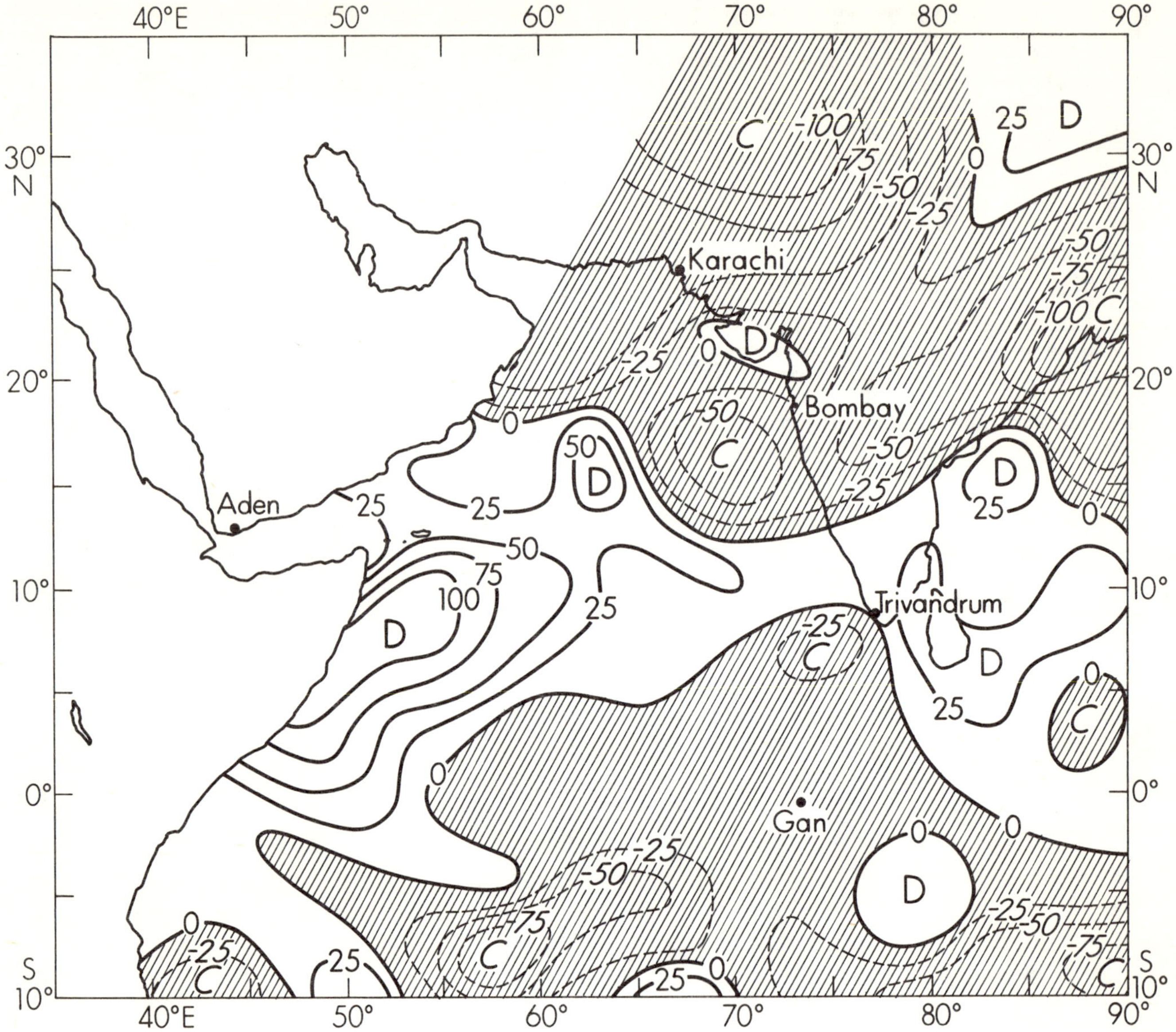

Figure 7. Divergence (10^{-7} sec^{-1}) of mean resultant low-levels for July.
Over sea -- from surface winds
(van Duijnen Montijn, 1952; Hantel, 1971)
Over land -- from winds at 900 m
(Miller and Keshavamurthy, 1968)
Centres of divergence denoted by "D";
centres of convergence denoted by "C".

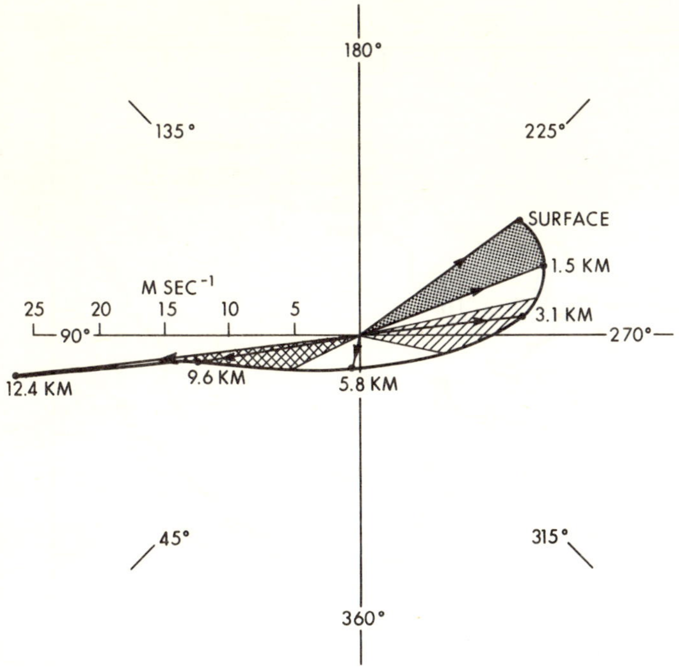

Figure 8. Change with height of mean resultant winds for July over the central Arabian Sea (10N, 60E). Primary origin of the air: South Indian Ocean: stippled, Africa: hatched; India, crosshatched.

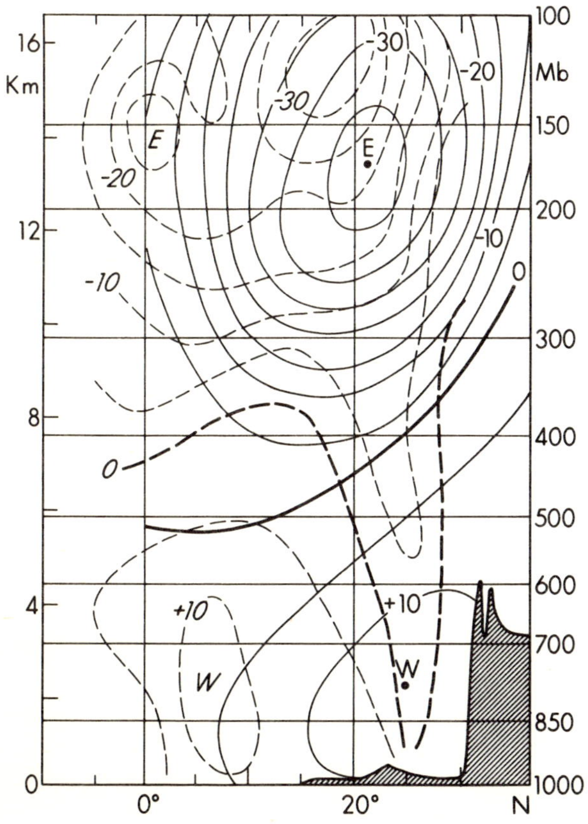

Figure 9. Zonal wind in metres per second for July along 80E.
Full lines -- computed after an 80-day integration (Murakami *et al.*, 1969).
Dashed lines -- from long-period radar-wind observations made at seven stations.

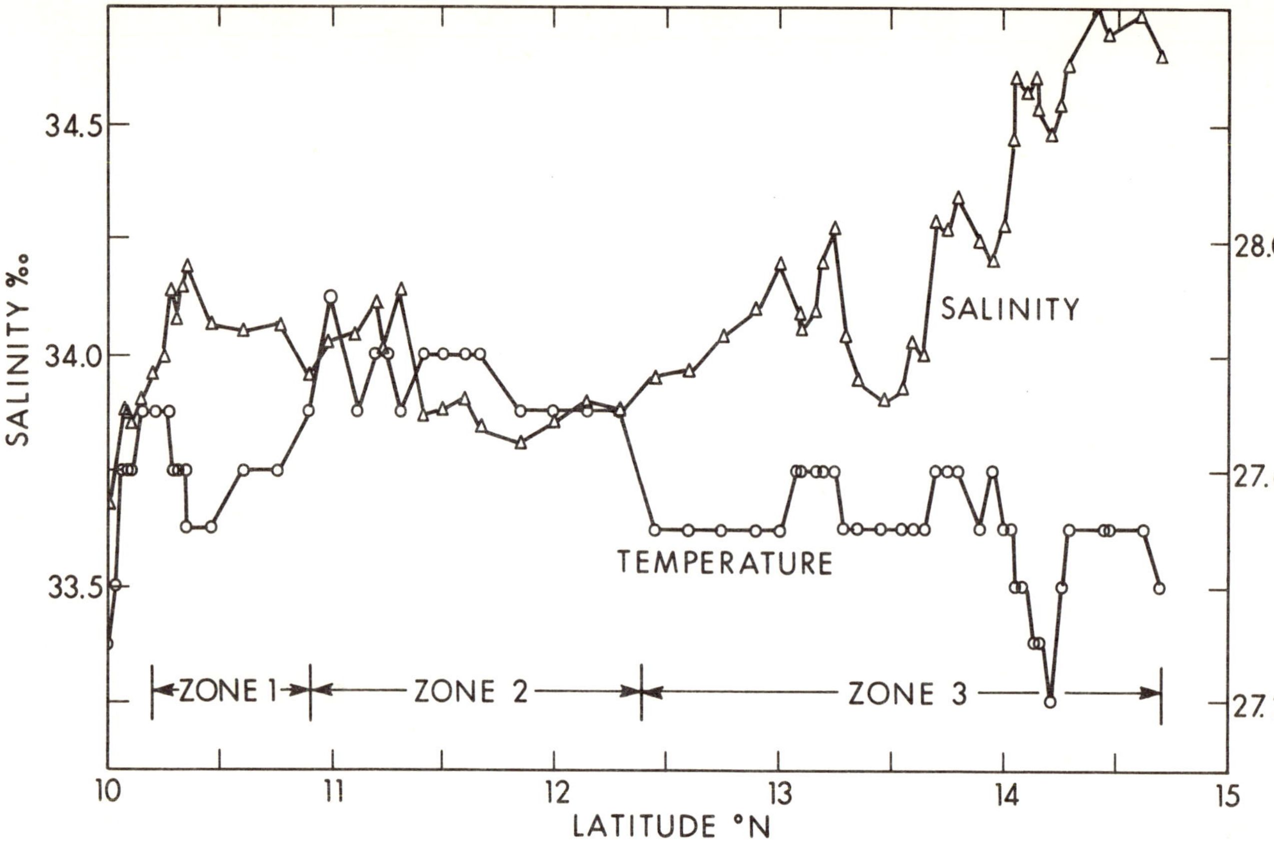

Figure 12. Significant temperature (°C) and salinity (‰) data points extracted from continuous surface measurements made on board Meteor on a cruise along the west coast of India from 14 to 15 February 1965. Average distance offshore: 100km.

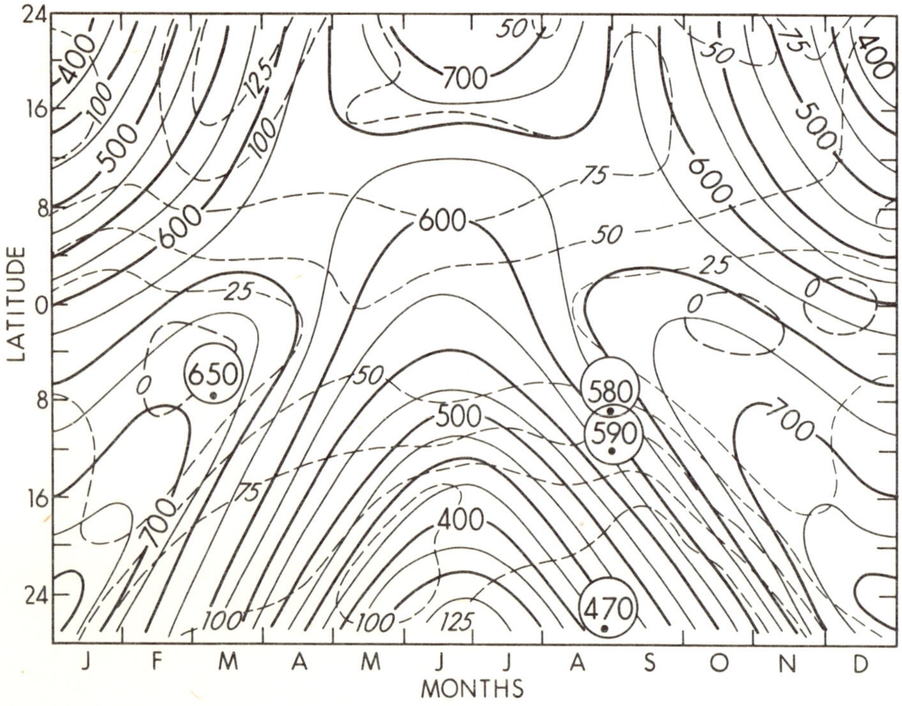

Figure 13. Latitudinal distribution of total incident solar radiation (langleys day^{-1} for cloudless day.
Full lines -- determined from measurements made at seven Indian Ocean area stations during 1963 and 1964 (Portman and Ryznar, 1971).
Dashed lines -- Budyko's (1956) values minus Portman and Ryznar's values.
Circled spot values -- determined from easter Indian Ocean shipboard measurements made during 1962 (Deacon and Stevenson, 1968).

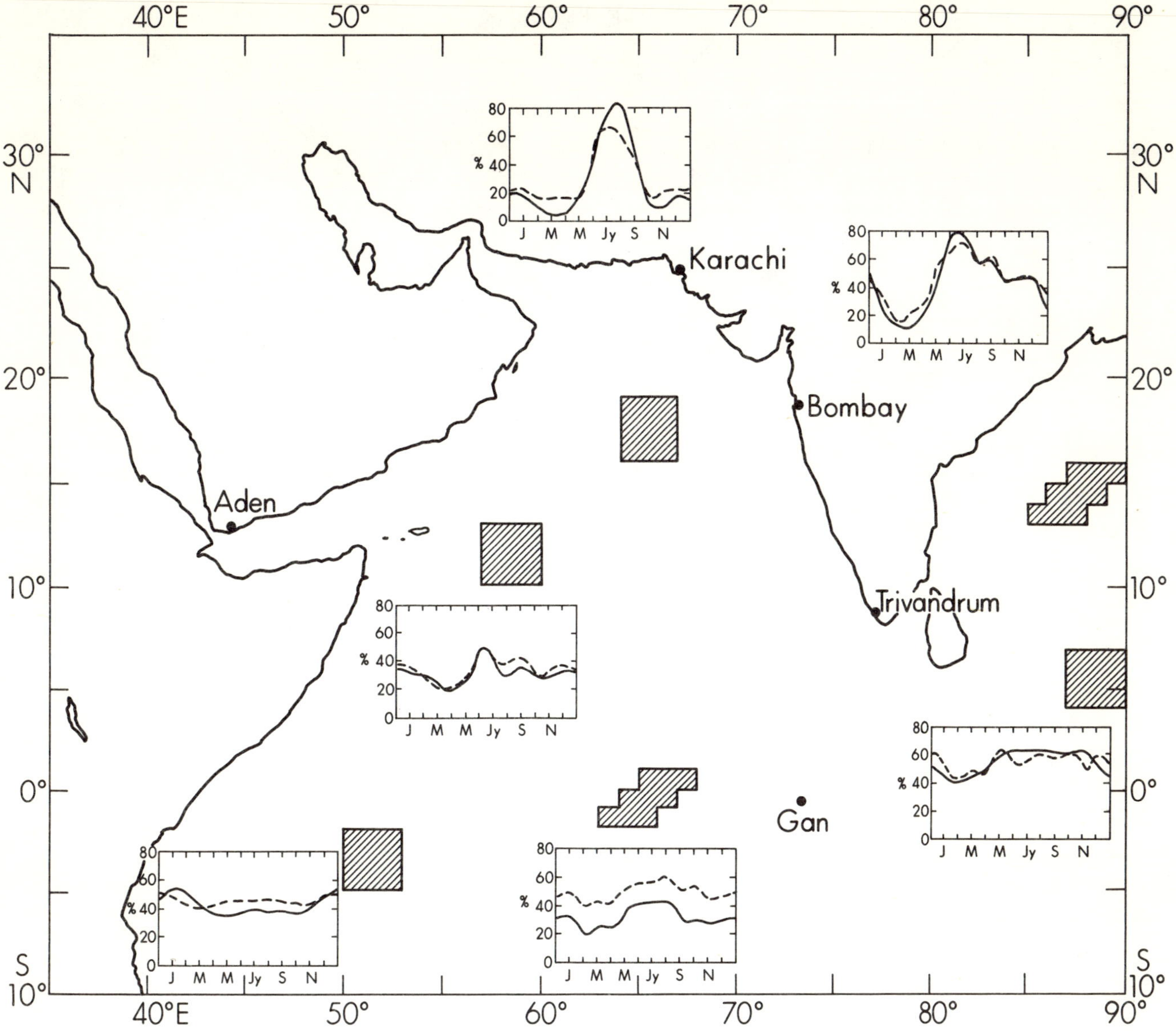

Figure 14. Mean monthly cloudiness (per cent) for six areas (hatched).
Full lines -- from long-period ship observations (U.S. Weather Bureau, 1957)
Dashed lines -- from 5 years of weather records (Sadler and Harris, 1970).

REFERENCES

BADGLEY, F.I., C.A. PAULSON, and M. MIYAKE, 1971: Profiles of wind, temperature and humidity over the Arabian Sea. International Indian Ocean Expedition, Meteorological Monographs, 6 (in press).

BUDYKO, M.L., 1956: The heat balance of the earth's surface, 256 pp. Gidrometeorologicheskoe Izdatel'stvo, Leningrad. /English translation (1958) U.S. Weather Bureau, Washington, D.C./.

BUNKER, A.F., 1967a: Interaction of the summer monsoon air with the Arabian Sea. In Proceedings Symposium Meteorological Results International Indian Ocean Expedition (P.R. Pisharoty, ed.), pp. 3-16. India Meteorological Department, New Delhi.

BUNKER, A.F., 1967b: Cloud formations leeward of India during the north east monsoon. Journal of Atmospheric Sciences, 24, 497-507.

BUNKER, A.F., and M. CHAFFEE, 1970: Tropical Indian Ocean clouds. International Indian Ocean Expedition, Meteorological Monographs, 4.

DEACON, E.L., and J. STEVENSON, 1968: Radiation and associated observations made on Indian Ocean cruises. I: Cruises during 1962. Melbourne, Australia, Division of Meteorological Physics, CSIRO, Technical Paper No. 16, 22 pp.

DESAI, B.N., 1967: The summer atmospheric circulation over the Arabian Sea. Journal of Atmospheric Sciences, 24, 216-220.

FLEAGLE, R.G., F.I. BADGLEY, and Y. HSEUH, 1967: Calculation of turbulent fluxes by integral methods, Journal of Atmospheric Sciences, 24, 356-373.

FLOHN, H., 1970: Local wind systems. In H. Flohn, ed. General Climatology. (Vol. 2 of World Survey of Climatology), pp. 139-171. Elsevier, Amsterdam, New York.

FLOHN, H., M. HANTEL, and E. RUPRECHT, 1968: Air mass dynamics or subsidence processes in the Arabian Sea summer monsoon? Journal of Atmospheric Sciences, pp. 527-529.

FUQUAY, D., and K. BUETTNER, 1957: Laboratory investigation of some characteristics of the Eppley pyrheliometer. Transactions of the American Geophysical Union, 38: 38-43.

HANTEL, M., 1971: Monthly charts of surface wind vergence over the tropical Indian Ocean. Bonner Meteorologische Abhandlungen, No. 14.

Intergovernmental Oceanographic Commission, 1965: Fourth Session: report of ad hoc IIOE coordination group. Unesco, Paris.

MILLER, F.R., and R.N. KESHAVAMURTHY, 1968: Structure of an Arabian Sea summer monsoon system. International Indian Ocean Expedition, Meteorological Monographs, 1.

MONAHAN, E.C., 1971: Oceanic whitecaps. Journal of Physical Oceanography, 1: 139-144.

MURAKAMI, T., R.V. GODBOLE, and R.R. KELKAR, 1969: Numerical simulation of the monsoon along 80°E. In Proceedings Conference Summer Monsoon South East Asia (C.S. Ramage, ed.), pp. 39-51. Navy Weather Research Facility, Norfolk, Virginia.

PORTMAN, D.J., and E. RYZNAR, 1971: An investigation of heat exchange. International Indian Ocean Expedition, Meteorological Monographs, 5.

RAMA, 1968: An attempt to trace the monsoon flow using natural radon. Indian Journal of Meteorology and Geophysics, 19, 167-170.

RAMAGE, C.S., 1968: Problems of a monsoon ocean. Weather, 23, 28-37.

RAMAGE, C.S., 1969: The setting of the south east Asian summer monsoon. In Proceedings Conference Summer Monsoon South East Asia (C.S. Ramage ed.), pp. 3-11. Navy Weather Research Facility, Norfolk, Virginia.

RAMAGE, C.S., F.R. MILLER, and C. JEFFERIES, 1971: International Indian Ocean Expedition, Meteorological Atlas, Vol. 1, The Surface Climate of 1963 and 1964. National Science Foundation, Washington, D.C.

RAMAGE, C.S., and C.R.V. RAMAN, 1971: International Indian Ocean Expedition, Meteorological Atlas, Vol. 2, Upper Air. National Science Foundation, Washington, D.C.

RODEN, G.I., 1961: On the wind-driven circulation in the Gulf of Tehuantepec and its effects upon surface temperatures. In Geofisica Internacional, pp. 55-76. Union Geofisica Mexicana, Mexico.

SADLER, J.C., 1969: Average cloudiness in the tropics from satellite observations. International Indian Ocean Expedition, Meteorological Monographs, 2.

SADLER, J.C., and B.E. HARRIS, 1970: The mean tropospheric circulation and cloudiness over South East Asia and neighboring areas. Hawaii Institute of Geophysics, Rep. No. 70-26.

SVERDRUP, H.U., 1951: Evaporation from the oceans. Compendium of Meteorology, 1071-1081.

U.S. Weather Bureau, 1957: U.S. Navy Marine Climatic Atlas of the World. Volume III, Indian Ocean.' Chief of Naval Operations, Washington, D.C., 7 pp., 267 charts.

VAN DIJK, W., 1956: An investigation of the vergence field of the wind and ocean currents of the Indian Ocean. Archiv für Meteorologie, Geophysik and Bioclimatologie, Serie A, 9, 158-177.

VAN DUIJNEN MONTIJN, J.A., 1952: Indian Ocean oceanographic and meteorological data, 2nd ed. Royal Netherlands Meteorological Institute de Bilt, No. 135.

WÜST, G., 1960: Proposed International Indian Oceanographic Expedition, 1962-1963. Deep Sea Research, 6, 245-249.

Table 3

TEMPERATURE AND SALINITY IN THE THREE ZONES SHOWN IN FIGURE 6

Zone number	1	2	3
Latitude range (°N)	10.2 - 10.9	10.9 - 12.4	12.4 - 14.7
Average temperature (°C)	27.6	27.8	27.5
Average salinity (°/₀₀)	34.07	33.94	34.29

Table 4

TOTAL SOLAR RADIATION ON CLEAR DAYS

(a) Measured on Australian cruises to the eastern Indian Ocean during 1962 (Deacon and Stevenson, 1968);
(b) From measurements made at Indian Ocean stations during 1963 and 1964 (Portman and Ryznar, 1971);
(c) From analysis of world-wide measurements (Budyko, 1956)

Mean latitude (s)	7°53'	8°41'	12°	27°12'
Mean date	9 March	1 September	1 September	28 August
Total solar radiation (ly day^{-1})				
(a)	650	580	590	470
(b)	680	615	585	435*
(c)	700	670	665	600

* extrapolated

Table 5

COMPARISON OF MEASURED (Badgley et al., 1971) AND ESTIMATED EVAPORATIVE COOLING OVER THE NORTH EASTERN INDIAN OCEAN IN FEBRUARY-MARCH 1964

Period	Location	No. of observations averaged	Sea temp. (°C)	Saturation vapour pressure at sea temp. (mb) e_s	Measurements at 8 m above sea surface: Vapour pressure (mb) e_a	Wind speed (m sec^{-1}) V	Evaporational cooling (ly day^{-1})	Calculated evaporational cooling* (ly day^{-1})
22-27 Feb.	18.5°N; 72°E	12	25.0	31.7	18.5	5.1	275	352
28 Feb. - 3 March	20°N; 69.5°E	47	25.3	32.2	20.6	5.3	304	318
4-6 March	19°N; 72E	26	24.6	30.9	21.4	5.8	234	285
7-9 March	20.5°N; 71.5°E	25	23.0	28.1	21.1	4.9	176	177

* Evaporational cooling = 5.17 (e_s - e_a) V - based on Sverdrup (1951).

A REVIEW OF RECENT PHYSICAL OCEANOGRAPHIC WORK IN THE INDIAN OCEAN

J. C. Swallow

National Institute of Oceanography
Wormley, Godalming, Surrey,
England.

INTRODUCTION

It is six years now since the formal end of the sea-going phase of the International Indian Ocean Expedition, but the analysis of its results has gone on actively, and much new work, both observational and theoretical, has been stimulated. In this paper I propose to review some aspects of these developments, particularly those of the last two years. Some new techniques of observation have come into more common use since 1965, and have not been much applied yet to the Indian Ocean. These too will be briefly reviewed, and I shall try to indicate what prospects there are for applying them to the physical situation now believed to exist in the Indian Ocean.

This review is by no means complete, and I apologize in advance for any glaring omissions. First, I shall try to answer some of the criticisms that used to be heard about the ineffectiveness of the Indian Ocean Expedition, and how it appeared to have done nothing about the obvious question of the phase relationship between the monsoon winds and the surface currents. That would have been a particularly difficult problem to tackle, within the framework of a large-scale oceanographic survey. Next, some new developments will be described that have clarified that problem considerably, at least as regards the onset of the Somali current. Then, because of its great general interest, recent work on upwelling in the Indian Ocean will be reviewed. Finally, because it was one of Professor Bruun's special interests, some recent work on the deep circulation will be mentioned.

THE MONSOON CIRCULATION

The most substantial continuing work on the IIOE material has been the preparation of the physical oceanographic atlas by Professor K. Wyrtki. This has necessarily involved much study of the data in order to decide how best to construct the atlas itself. There are three distinct large-scale circulation systems; from south to north these are the Indian Ocean sector of the Circumpolar Current, the sub-tropical anticyclonic gyre, and the seasonally changing monsoon gyre. The latter two are separated by a well-defined front near 10°S. Unfortunately this review had to be prepared just before the expected date of publication of the atlas. How ever, a preview of some of the results compiled for the monsoon gyre has been given by Düing (1970), in the form of charts of the dynamic topography of the sea surface relative to 1,000 decibars for five periods of the year. He had hoped to work with monthly charts, but the data were too sparse, even though the total number of hydrographic stations available was much greater than had been used in the past for similar studies in other oceans.

For example, Defant (1941) used 629 stations in his charts of the dynamic topography of the Atlantic, and Reid (1961) had 1,729 stations for the whole of the Pacific Ocean. By contrast, Düing used 3,060 stations for the much smaller area of the Indian Ocean north of 20°S. And they were not very unevenly distributed in time either. They are inadequate because of the complicated situation that is now seen to exist in the northern part of the Indian Ocean.

The monsoon gyre appears to contain many relatively strong cyclonic and anticyclonic eddies, with dimensions in the range 100 to 1,000 km., capable of changing dramatically in 2 months or less. They seem to be more pronounced in comparison to the large-scale circulation in this northern part of the Indian Ocean than in the same latitude zones of the other oceans, if the contoured charts of surface dynamic height anomaly can be taken as a guide. It might be suspected that they are related in some way to the strong boundary current (or possibly to the larger number of stations used), but there seems to be little doubt about the reality of these relatively strong, medium-scale eddies. Their presence sets a limit to the accuracy with which the larger-scale circulation can be defined from the available observations, and makes it seem unlikely that conventional surveys could be made intensively enough to determine the phase of the overall response of the ocean to the monsoon with useful accuracy.

Another approach has been more successful; various mathematical models have been developed from which predictions about the ocean's response have been made, and these have been compared with observations inferred from sea surface temperatures from satellites. Düing (1970), impressed by the complex pattern of vortices observed, has produced a model in which such features can be generated, using a periodic wind stress simulating the monsoon. By choosing a sufficiently small value of a parameter representing dissipation, he could reproduce the observed transport and width of the Somali Current, and a phase lag of no more than 10 days was predicted for its onset. However, even though several observed features seem to be realistically reproduced, the neglect of stratification and unusually low dissipation needed are serious disadvantages.

A more realistic model has been formulated by Cox (1970), a 3-dimensional numerical model with 7 layers and a minimum grid size of 1° square, covering the Indian Ocean north of 20°S. The monsoon was simulated by a periodic wind stress and, in addition to the description of the overall response of the ocean, the process of establishment of the Somali Current was considered in detail. He concluded that the strong part of the current north of 3°N was mainly due to local winds causing an offshore Ekman transport, with consequent upwelling, tilting of shallow density surfaces, and associated geostrophic current. The main features of the Somali Current were qualitatively well reproduced in the model, though the estimated transport was about half, and width about twice, the observed maximum values. Because of the importance of the local wind in this case, the predicted phase lag between wind and current was small - less than the half-month averaging time used.

A model of quite a different kind was considered by Lighthill (1969). He worked out analytically the response of the ocean to a sudden application of negative wind stress curl (corresponding to the south west monsoon onset) along a zonal strip near the equator. In such a situation, transient current patterns would be set up, propagating a northward flux westwards and concentrating it into a western boundary current. Moreover, he showed that most of the transport in the upper layer would be in the baroclinic mode and that near the equator its rate of propagation would be quite high, much faster than that calculated by Veronis and Stommel (1956) for mid-latitudes. Without taking any winds within 500 km of the coast into account, Lighthill found that a transport equal to 40% of the observed maximum of the Somali Current could be built up by this process in one month.

Recent observations suggest that probably both the mechanisms put forward by Lighthill and by Cox are important in building up the Somali Current. Düing and Szekielda (1971), using synoptic charts of sea surface temperature obtained from infra-red observations from the Nimbus satellites, have traced the development of the strong horizontal temperature gradients off the Somali coast during three south west monsoon seasons and related them to the increase in wind speed. They used the temperature gradients as an index of the strength of the baroclinic part of the current, and showed that the early growth of the current was in phase with the increase in local south west wind, as suggested by Cox. About a month later, however, a further increase in temperature gradient occurred in advance of any local wind increase, indicating the arrival of that part of the current generated according to Lighthill's mechanism.

This satisfactory agreement between model predictions and observation does not of course mean that the whole problem of understanding the monsoon circulation is solved. Complex and ingenious as these models are, they are in a sense only first approximations. Each uses an idealized form of wind stress distribution to represent the actual onset of the monsoon.

Some recent current measurements by Leetmaa (1971), for example, show that south of the equator the start of the Somali Current was in phase with the local wind; but growth within the next few weeks came from the strengthening of the South Equatorial Current in response to the increasing south easterly monsoon. North of the equator, observations by Bruce and Volkmann (1969) suggest that the strong anticyclonic eddy known to exist offshore from the Somali Current in the south west monsoon (see, for example, Swallow and Bruce, 1966), persisted below the surface in the north east monsoon season, and that would doubtless have an effect on the re-growth of the Somali Current. There is only a slight indication of the presence of this strong eddy in Cox's numerical results, probably because in his model the eddy processes were simply represented by an eddy diffusion coefficient, chosen with the horizontal grid size in mind.

Admittedly this representation of eddies and their interactions may be inadequate, as the results of Ozmidov, Belyaev and Yampolsky (1970) suggest. From a two-month series of measurements with an anchored array of current meters in the north west Indian Ocean, they found that in the near-surface layers energy was being transferred from larger scales to the smaller fluctuations, but at depths below a few hundred metres the direction of energy transfer was reversed.

What are the prospects for future studies of the monsoon circulation? Clearly the situation is sufficiently complex and rapidly varying that anyone planning observations needs all the guidance he can get from the most realistic models available. Reciprocally these models will be improved by taking into account more of the observed features.

The importance of learning more about the medium-scale eddies in further studies of the Indian Ocean will be evident from what has been said above; there are formidable observational difficulties in dealing with the range of time scales and horizontal dimensions, and of course this need to understand the rôle of eddies is a general problem not confined to the Indian Ocean. One helpful feature is that, for at least some time scales, the currents often seem to have a relatively simple structure in the vertical (see, for example, Titov and Fomin, 1971).

As regards techniques of observation, new developments have appeared and others have come into more common use that could with advantage be applied to the study of the monsoon circulation. The synoptic charts already mentioned of sea surface temperature from the satellites with infra-red sensors are of course extremely effective in revealing rapid changes over large areas. It would be desirable though to have more confirmation of inferences made from them, for example simultaneous observations of surface current to show whether in the early stages of upwelling on the Somali coast the horizontal gradients of temperature really indicate a large-scale baroclinic current or not. Perhaps the air-dropped transport floats being developed by W. S. Richardson (an extension of the method described by Richardson, Carr and White, 1969) would be most suitable for such a purpose.

Many improvements have been made in the capabilities of ships making conventional oceanographic surveys since the time of the IIOE. For some ships, the uncertainties of navigation in strong currents have been reduced by use of world-wide navigation aids such as the U. S. Navy navigational satellite system. Dead reckoning has also been improved and surface currents are extracted more reliably. Various instruments giving continuous vertical profiles of temperature and salinity are used much more commonly now as a supplement to conventional water sampling, though some were already in use during the IIOE (Hamon, 1967). Thanks to these advances, research vessels could now produce much more accurate observations of the vertical distribution of density and velocity in the Somali Current region than the Argo and Discovery did in August 1964 (Swallow and Bruce, 1966). On the other hand, the speeds of ships and rate of working stations have not increased appreciably, and the overall description of the changing field of density and velocity might be only slightly improved. This does not mean, of course, that further oceanographic survey work there is pointless, only that such work would need to be planned with the smallness of horizontal scale and rapid rate of change clearly in mind, as well as the high speed of the current.

UPWELLING

Seasonal upwelling in the northern Indian Ocean is an important aspect of the monsoon circulation, closely related to the setting up of the horizontal currents as can be seen from the above-mentioned work of Cox (1970) and Düing and Szekielda (1971). Cox's model shows clearly the high vertical velocities in the upwelling region at the inshore edge of the northern part of the Somali Current - speeds of nearly 5 m per day in the top 200 m - which he points out could be produced by the local wind stress alone. A similar high vertical speed was inferred by Swallow and Bruce (1966) from the rate of rise of the 20°C isotherm in a series of sections downstream along the Somali Current. Within the limits of its resolution, Cox's model describes the upwelling region very well, and it is to be hoped that ways will be found to cope with the large amount of computing that must be needed in models with finer resolution. The sharp gradients of surface temperature reported by Warren et al. (1966) at the boundary of the cool upwelled water suggest that horizontal scales as small as 10 km may be significant in the Somali coast upwelling region.

It would be of interest also if model studies could be made of the upwelling resulting from different patterns of wind stress, and if the processes linking wind stress, small-scale upwelling, and generation of the geostrophic current could be modelled in more detail. The Somali coast upwelled water varies dramatically in extent in different years, and in the timing within each year, as can be seen in the surface temperature charts of Currie et al. (1971) and Düing and Szekielda (1971). The latter authors mention the difficulty of finding sufficiently representative wind observations in that area, to relate to the observed upwelling - for once, oceanographers seem to be more fortunate than meteorologists as regards observations - but only by courtesy of the Nimbus satellites.

The upwelling off the southern coast of Arabia as seen in the summer of 1963 has been discussed by Currie et al. (1971) and contrasted with the Somali coast upwelling. While the latter seemed to be closely related to the strong boundary current and its turning offshore, the Arabian coast upwelling occurred in a broad band along more than 1,000 km of coastline and was associated with relatively weak horizontal currents. Bottero (1969) calculated the vertical and horizontal transports in that region, using the method applied by Wyrtki (1963) to the Peru Current. He found that the upwelling extended at least 400 km offshore, and that there were upward velocities at all depths down to at least 700 m. These were strongest in a zone within 50 km of the coast, and were of the order of 10 m per day in the upper 100 m.

Some direct measurements of horizontal currents were made by the Discovery at the same time as the hydrographic survey on which Bottero's calculations were based; some of these were consistent with the calculated field of motion but others suggested that, as might be expected close inshore, local topographic effects were important. Detailed temperature sections from bathythermograph observations suggested that once again horizontal scales as small as 10 km could be important near the coast, and these were not fully resolved in the coarser grid of hydrographic stations. The weak upwelling in a broad zone offshore was sensitive to the manner of smoothing the wind observations from which the divergence of the Ekman transport was calculated, though there was a consistent increase in wind speed with distance from the coast. In view of the uncertainties involved, and the difficulty of estimating these uncertainties quantitatively, it seems desirable that further confirmation should be sought for the unusual width and depth of this upwelling zone off the Arabian coast. Such confirmation might be obtained by direct measurement of the deep vertical velocities, using a flock of the rotating floats developed by Webb, Dorson and Voorhis (1970). An array of moored current meters, with anemometers on surface buoys, closely spaced inshore, should be used together with a dense network of hydrographic observations to help clarify the pattern of horizontal transport and its fluctuations.

THE DEEP CIRCULATION

The model of the abyssal circulation initiated by Stommel (1958), with western boundary currents in the deep water of all the oceans, has gained further support recently and some of the evidence has come from the Indian Ocean. Kuo and Veronis (1970) have extended the work of Arons and Stommel (1967) by considering the combined effects of advection, diffusion and decay on the distribution of tracers in a two-dimensional numerical model of the abyssal world ocean. They showed that the observed distribution of dissolved oxygen at 4 km. depth could be modelled in surprisingly good detail in this way, with suitable choice of values for the intensity of circulation and eddy diffusivity. The north eastward decrease in the Indian Ocean came out as clearly as could be expected, with the simplified boundaries of this model. They look forward to the possibility of incorporating bottom topography and the effect of the thermocline in more detail. The deep observations being compiled in the Indian Ocean Atlas will be valuable for testing such future models. The sequence of deep basins with narrow connexions at 4, 000 m. depth on the western side of the Indian Ocean may be of particular interest for that purpose. On the question of the existence of a deep western boundary current in the Somali Basin, for which Warren et al. (1966) found some evidence from the salinity distribution, it may be noted that there is a slight indication of northward flow in the lowest layer of Cox's numerical model, under the Somali Current (Cox (1970), Fig. 14).

Warren (1971) has found more positive, but still indirect, evidence of a deep western boundary current off the eastern slope of Madagascar. In a section at 12°S, the slopes of the isotherms indicated relative northward flow below 3, 200 m. over a width of about 500 km, with maximum relative speeds of 6 to 7 cm/sec. Less definite evidence was found in a section at 23°S, and no direct measurements of current were made, so that although there was strong qualitative evidence for the deep boundary current it was not possible to estimate its transport reliably for comparison with the model.

Lal, Nijampurkar and Somayajulu (1970) have estimated the residence time of deep water in the Pacific and Indian Oceans from measurements of silicon-32 concentration. In fact, however, because of uncertainty in the rate of decay of silicon-32, they find it more useful to assume the deep water residence times indicated by radiocarbon measurements and use their results to get a better estimate of the half-life of silicon-32. What makes it seem particularly appropriate to mention this work here is that the measurements were made, not on samples of deep water, but on parts of siliceous sponges dredged by the Galathea.

Future work on the deep circulation of the Indian Ocean could be expected to include long-term current measurements using anchored current meters, to give some direct evidence about the existence of a deep western boundary current and the flow between deep basins. The greater resolution, particularly in temperature, attainable with STD recorders could be usefully applied to some situations in the deep water, in future hydrographic surveys.

Perhaps it is premature to comment before the Indian Ocean Atlas has been appreciated, but it seems to me that not much use was made of the reference stations designated at an early stage of the IIOE. Rochford (1965) pointed out that many of them would be useless for intercalibration purposes but might instead yield information about fluctuations in time. However, I doubt whether they were occupied often enough. For the deep water, it might have been better to use some of the small isolated basins as reference stations. In one of these, near 9°S, $67^1/2$°E and 6, 400 m. deep, the Discovery found uniform salinity and potential temperature below 4, 000 m. The Galathea station in the Sunda trench showed similar uniformity (but different values, of course) below 4, 800 m. Numerous other small holes exist and probably most of them have not been sampled; they could serve as useful checkpoints for intercalibration. On the other hand, they could preserve evidence of changes that might still be detected even with very intermittent monitoring. Some of them might even contain hot brines, like the ones in the Red Sea. Working stations in such situations may not be particularly easy, but it could be useful. It seems like a project that might have appealed to Professor Bruun.

REFERENCES

ARONS, A. B., and Henry STOMMEL (1967)
On the abyssal circulation of the World Ocean - III. An advection-lateral mixing model of a tracer property in an ocean basin. Deep-Sea Research, Vol. 14. 441-457.

BOTTERO, J. S. (1969)
An analysis of upwelling off the south east Arabian coast during the summer monsoon. M. S. thesis, Oregon State University. 70 pp. (Unpublished manuscript).

BRUCE, J. G., and G. H. VOLKMANN (1969)
Some measurements of current off the Somali coast during the north east monsoon. Journal of geophysical research, 74. 1958-1967.

COX, Michael D. (1970)
A mathematical model of the Indian Ocean. Deep-Sea Research, 17: 47-75.

CURRIE, R. I., A. E. FISHER and P. M. HARGREAVES (1971)
Arabian sea upwelling. Symposium on the biology of the Indian Ocean, Kiel, 31 March - 6 April 1971 (in press)

DEFANT, A. (1941)
Die absolute Topographie des physikalischen Meeresniveaus und der Druckflächen, sowie der Wasserbewegungen im Atlantischen Ozean. Deutsche Atlantische Expedition "Meteor" 1925-1927, Wissenschaftliche Ergebnisse, Bd. VI. s. 191-260.

DÜING, Walter (1970)
The Monsoon régime of the currents in the Indian Ocean. East-West Center Press, Honolulu. 68 pp.

DÜING, Walter, and Karl-Heinz SZEKIELDA (1971)
Monsoonal response in the Western Indian Ocean. Journal of geophysical research, 76: 4181-4187.

HAMON, B. V. (1967)
Medium-scale temperature and salinity structure in the upper 1, 500 m. in the Indian Ocean. Deep-sea research, 14: 169-181.

KUO, Han-Hsiung, and George VERONIS (1970)
Distribution of tracers in the deep oceans of the world. Deep-sea research, 17: 29-46.

LAL, D., V. N. NIJAMPURKAR and B. L. K. SOMAYAJULU (1970)
Concentrations of cosmogenic Si32 in Deep Pacific and Indian waters based on study of Galathea deep sea siliceous sponges. Galathea report, 11: 247-256.

LEETMA, Ants (1971)
The response of the Somali Current to the south west monsoon of 1970. Deep-sea research (in press)

LIGHTHILL, M. J. (1969)
Dynamic response of the Indian Ocean to onset of the south west monsoon. Philosophical transactions of the Royal Society of London, pp. 45-92.

OZMIDOV, R. V., V. S. BELYAYEV and A. D. YAMPOLSKIY (1970)
Some features of the turbulent energy transfer and the transformation of turbulent energy in the ocean. Izvestiya, Academy of Sciences, USSR atmospheric and oceanic physics, Vol. 6, English edition translated and published by American Geophysical Union, pp. 160-163.